History of Physics

The Springer book series *History of Physics* publishes scholarly yet widely accessible books on all aspects of the history of physics. These cover the history and evolution of ideas and techniques, pioneers and their contributions, institutional history, as well as the interactions between physics research and society. Also included in the scope of the series are key historical works that are published or translated for the first time, or republished with annotation and analysis.

As a whole, the series helps to demonstrate the key role of physics in shaping the modern world, as well as revealing the often meandering path that led to our current understanding of physics and the cosmos. It upholds the notion expressed by Gerald Holton that "science should treasure its history, that historical scholarship should treasure science, and that the full understanding of each is deficient without the other." The series welcomes equally works by historians of science and contributions from practicing physicists.

These books are aimed primarily at researchers and students in the sciences, history of science, and science studies; but they also provide stimulating reading for philosophers, sociologists and a broader public eager to discover how physics research – and the laws of physics themselves – came to be what they are today.

All publications in the series are peer reviewed. Titles are published as both print- and eBooks. Proposals for publication should be submitted to Dr. Angela Lahee (angela.lahee@springer.com) or one of the series editors.

More information about this series at https://link.springer.com/bookseries/16664

Michael Eckert

Turbulence—an Odyssey

Origins and Evolution of a Research Field at the Interface of Science and Engineering

Springer

Michael Eckert
Forschungsinstitut
Deutsches Museum
Munich, Germany

ISSN 2730-7549 ISSN 2730-7557 (electronic)
History of Physics
ISBN 978-3-030-91458-5 ISBN 978-3-030-91459-2 (eBook)
https://doi.org/10.1007/978-3-030-91459-2

This Springer imprint is published by the registered company Springer Nature Switzerland AG
The registered company address is: Gewerbestrasse 11, 6330 Cham, Switzerland

Odyssey: a long trip or period involving a lot of different and exciting activities, esp. while searching for something.

https://dictionary.cambridge.org/dictionary/english/odyssey

Preface

Turbulence is a research field where high expectations are met with recurrent frustration. "Turbulence is the most important unsolved problem of classical physics," Nobel Laureate Richard Feynman (1918–1988) is quoted as one among other renowned scientists who voiced the riddle of turbulence. The famous author of a classic textbook on *Hydrodynamics*, Horace Lamb (1849–1934), is alleged to have said "When I meet my Creator, one of the first things I shall ask of Him is to reveal to me the solution to the problem of turbulence" (Peter A. Davidson, 2011, p. xiii). Theodore von Kármán (1881–1963), another pioneer of this science, put a similar quote in the mouth of Arnold Sommerfeld (1868–1951), the famous atomic physicist. Sommerfeld "once told me," Kármán recalled, "that before he died he would like to understand two phenomena–quantum mechanics and turbulence. Sommerfeld died in 1924. I believe he was somewhat nearer to an understanding of the quantum, the discovery that led to modern physics, but no closer to the meaning of turbulence." (von Kármán, 1967, p. 134). Recollections of this sort lack historical scrutiny and authenticity (Sommerfeld died in 1951, not in 1924), but they highlight a common perception among physicists, mathematicians and engineers: that there is a big mystery behind the phenomenon of turbulence.

There are different approaches towards a history of turbulence. From the vantage point of "the turbulence problem," I have traced the changing perception of what was regarded as the outstanding riddle in response to the research agendas in different environments and at different times (Eckert, 2019b). Another approach followed the outstanding pioneers of this research field (Peter A. Davidson, 2011). In this book, I sketch the history of turbulence in a different manner—as an odyssey through uncharted territories from the seventeenth century until a crucial stage in the mid of the twentieth century, a conference at Marseille in the year 1961. This event marked the inauguration of turbulence as "a new science" (Favre, 1962, p. 5), a perception that was confirmed by the subsequent developments and celebrated half a century later with another memorable conference at Marseille (Farge, 2013).

The portrayal of turbulence as an odyssey yields a less coherent view of its origins. There is room for contingency and haphazard developments within and beyond various scientific and engineering disciplines that were themselves in the making when they are observed over longer periods of time. Yet the narrative of such a history must not result in a chaotic sequence of events: I structure this odyssey in three parts that explore the gateways (Part I) towards turbulence prior to the early twentieth century, the formation of mayor concepts (Part II) until World War II and the staging of the "new science" of turbulence prior to Marseille '61 (Part III). Besides some observations in the epilogue, I do not venture to review turbulence beyond 1961. Such a history would have to include in particular the rise of computational fluid dynamics, turbulence modelling (the "closure problem"), nonlinear dynamical systems ("strange attractors") and unresolved mathematics of the Navier–Stokes equation ("blow-up")—plenty of challenges for a future study of what appears to be another odyssey.

Munich, Germany
September 2021

Michael Eckert

Acknowledgements

I am grateful to the staff of the archives and libraries cited in the footnotes for their help to find the pertinent source material and to my colleagues at the Deutsches Museum where I found the appropriate environment for work of this nature. This book benefited from previous projects at the Deutsches Museum on "Physik zwischen naturwissenschaftlich und technologisch orientierter Forschung: Das Beispiel der Strömungsforschung" and "Ludwig Prandtl—Wissenschaftler, Forschungsmanager und Politikberater," supported by the Deutsche Forschungsgemeinschaft (FOR 393/2-1, FOR 393/2-2 and TR 364/18-1). My thanks go also to the organizers of conferences that provided me opportunities to present one or another part of this work, in particular the International Colloquium 2011 in Marseilles on "Fundamental Problems of Turbulence: 50 years after the Turbulence Colloquium Marseilles 1961."

Contents

Part II Concepts

Part III Staging Turbulence

Acronyms

APS	American Physical Society
AVA	Aerodynamische Versuchsanstalt, Göttingen
DFD	Division of Fluid Dynamics
DFDA	Division of Fluid Dynamics, Archives
DLR	Deutsches Zentrum für Luft- und Raumfahrt
DMA	Deutsches Museum, Archiv, Munich
DMV	Deutsche Mathematiker-Vereinigung
EUROMECH	European Mechanics Society
FIAT	Field Intelligence Agency, Technical
GALCIT	Guggenheim Aeronautical Laboratory at the California Institute of Technology, Pasadena
GAMM	Gesellschaft für Angewandte Mathematik und Mechanik
GOAR	Göttingen Archive of the DLR
IAS	Institute of the Aeronautical Sciences
IAU	International Astronomical Union
ICSU	International Council of Scientific Unions
IGY	International Geophysical Year
ISMT	Institute for the Statistical Mechanics of Turbulence
IUGG	International Union of Geodesy and Geophysics
IUTAM	International Union of Theoretical and Applied Mechanics
LFA	Luftfahrtforschungsanstalt Hermann Göring
MPGA	Max-Planck-Gesellschaft, Archiv, Berlin
NACA	National Advisory Committee for Aeronautics
NOL	Naval Ordnance Laboratory
NPL	National Physical Laboratory, Teddington
ONERA	Office National d'Études et de Recherches Aérospatiales
SUB	Staats- und Universitätsbibliothek, Göttingen
TKC	Theodore von Kármán Collection, California Institute of Technology, Pasadena
UAM	University Archive Munich
UNESCO	United Nations Educational, Scientific and Cultural Organization

ZAMG	Zentralanstalt für Meteorologie und Geodynamik, Vienna
ZAMM	Zeitschrift für Angewandte Mathematik und Mechanik
ZWB	Zentrale für Wissenschaftliches Berichtswesen über Luftfahrtforschung

Part I
Roots

Turbulence did not emerge as a specialty by branching off from physics or another scientific or engineering discipline. Its roots have grown since the early modern era in a host of fields, from hydraulics to meteorology. This part traces the gateways to turbulence until the early decades of the 20th century. Only then was turbulence discerned as a challenge that called for a concerted effort beyond disciplinary boundaries.

Chapter 1
Pipe Flow

Abstract Pipe flow has been a challenge that gave rise to investigations on turbulence—long before turbulence was discerned as a research problem in its own right. The discharge of water from elevated reservoirs through long conduits such as for the fountains at Versailles suggested investigations about the resistance in relation to the different diameters and lengths of the pipes as well as the speed of flow. Despite numerous measurements the data could not be reproduced by a commonly accepted formula, not to mention a theoretical derivation. Even more inaccessible to rational elaboration was the resistance of air flow in long pipes for the supply of blast furnaces or mine air. In the 19th century it became gradually clear that there were two modes of pipe flow, laminar and turbulent. While the former could be accommodated under the roof of hydrodynamic theory, the latter proved elusive.

1.1 The Conduits at Versailles

Conduits of pipes were central parts of water technology since antiquity. Roman pipes even seem to have been standardized. By the middle ages a variety of pipes was in use, made from wood, lead or ceramic. Long conduits were composed as a rule by pipes of ten to twelve feet length. Longer pipes would have been too cumbersome to handle, shorter pipes would have required more joints with the result of a higher friction. But this practice did not result from flow measurements. The design and construction of medieval pipes was entirely based on practical experience (Magnusson, 2001, Chap. 3).

The first published record of pipe flow measurements concerned the requirements for the water art at Versailles and other royal gardens of the "Sun King", Louis XIV., such as the height of fountain jets driven from the discharge of elevated water reservoirs. Under these premises, hydraulics was put on the agenda of the newly founded royal academy in Paris (Blay, 1986). Two years after its foundation, on 11 July 1668, the academicians Jean Picard (1620–1682) and Edme Mariotte (1620–1684) were charged with a study of "the force of running water for pressing and

M. Eckert, *Turbulence—an Odyssey*, History of Physics,
https://doi.org/10.1007/978-3-030-91459-2_1

moving",[1] a task that led to a number of experimental investigations on the relation of water discharge through pipes from reservoirs at different heights and through circular holes of various diameter. The results were published in academy memoirs and summarized in a *Traité du Mouvement des Eaux et des Autres Corps Fluides* (Mariotte, 1686). Not least because of this work, published posthumously by his academic colleague Philippe de La Hire (1640–1718), Mariotte is regarded as "the father of the experimental method in France" (Rouse and Ince, 1957, p. 63).

Pipe flow measurements at Versailles were not a one-time task, nor were Mariotte and Picard the only academicians concerned with this investigation. In the1660s and1670s Versailles and its water supply system was a giant construction site under permanent academic surveillance with regard to the levelling of the canals, pipes and reservoirs for the fountains in the park (Descamps, 2003). It involved outstanding scholars such as Christiaan Huygens (1629–1695) and Ole Rømer (1644–1710) as well as less known academicians like Claude-Antoine Couplet (1642–1722), for whom Versailles was an opportunity to develop his hydraulic skills with pipe flow measurements until he rose to prominence as the academy's permanent treasurer (de Fontenelle, 1722). In 1696 Couplet's son, Pierre (1670–1743), also became a member of the academy—and a collaborator of his father on the pipe flow experiments at Versailles (Heyman, 1976). Many years later, on 22 March 1732, he presented to the academy a comprehensive report about these investigations (Couplet, 1732). The "rules" of pipe flow, as reported in Mariotte's treatise, were derived from measurements on very short conduits, Couplet junior criticized the earliest pipe flow measurements at Versailles. The crucial question concerned the resistance of the water in the conduits:

> This is a question which despite its importance has hardly been clarified; its solution seems to depend on the knowledge of the friction of the water in the pipes in relation to the different diameters, lengths and speed of the water.[2]

Couplet argued that only a great number of experiments could provide this knowledge, and that Versailles was the perfect site for these measurements. By the early 18th century water supply systems did not yet involve large pipes made from cast iron like those at Versailles, where only the best was good enough to achieve the King's demands. Couplet's experiments focused on the discharge of water through five conduites made of cast iron pipes that varied in length from 296 to 2340 toises (1 toise = 6 ft = 1.83 m) and in diameter from 4 to 18 pouces (1 pouce = 1 in. = 2.54 cm). Through these long pipes the discharge was quite different from what was to be expected according to Mariotte's rules. Couplet attributed this discrepancy to the friction of the water against the inner wall of the pipes:

---

[1] "de la force des eaux courantes à presser et à mouvoir." (Blay, 1986, p. 93).

[2] "C'est une question qui, toute importante qu'elle est, n'a point encore été éclaircie, et dont il paroît que la solution dépend de la connoissance du frottement des eaux dans leurs conduites par rapport à différents diamètres et longueurs différentes, et par rapport à leurs différentes vitesses." (Couplet, 1732, p. 114).

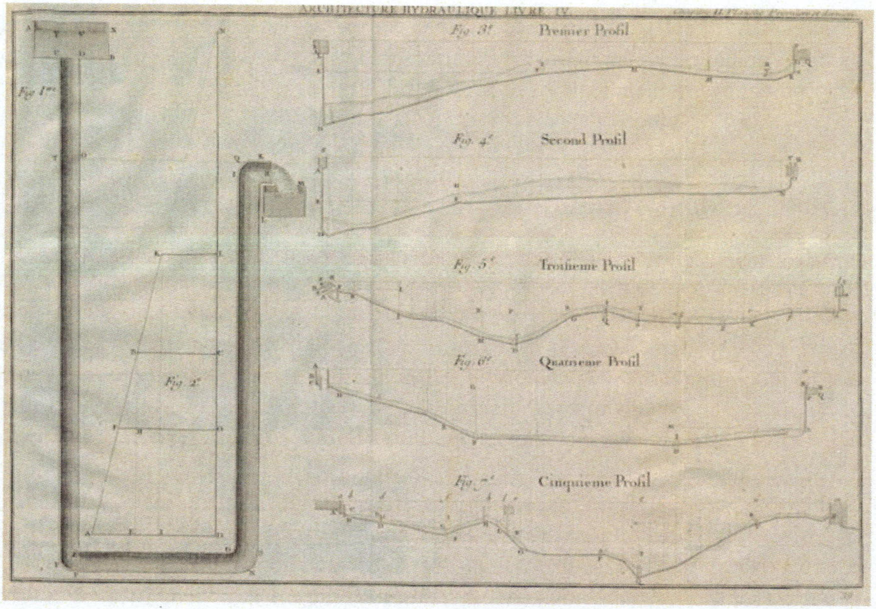

Fig. 1.1 The five conduits at Versailles used for Couplet's pipe flow experiments (Belidor, 1737, Livre IV, Chapitre II, plate I)

> ... and it is astonishing that this friction of the water against the walls of this pipe caused a decrease in the flow of water about 30 times greater than the quantity of water which came out through this pipe.[3]

The conduits included elbows where trapped air disrupted the water column in the pipes. In the case of the longest conduit it could take more than ten days until the water from the reservoir was discharged. Couplet did not arrive at a formula for pipe flow friction, but the conclusion from his measurements was obvious: Friction retarded the flow of water in the pipes considerably—the smaller the pipe diameter and the longer the conduit and the larger the flow speed, the more. Elbows and bends, too, added to the friction.

Couplet's measurements were regarded important enough to be included five years later in Bernard Forest de Bélidor's (1697–1761) famous *Architecture hydraulique* (Belidor, 1737, Livre IV, Chapitre II). They deserved particular interest also because of their location at Versailles, "perhaps the only place of the world where one finds everything one could wish in order to perform experiments of that sort we are dis-

[3] "... il est étonnant que ce frottement de l'eau contre les parois de ce tuyau ait causé une diminution d'écoulement d'eau environ 30 fois plus grande que la quantité d'eau qui est sortie par cette conduite." (Couplet, 1732, p. 148).

cussing here."[4] He paid tribute to Couplet by extensive quotes from his academy memoir, including a copper engraving that showed the profile of the five conduits at which the measurements had been performed (Fig. 1.1).

1.2 Pipe Flow Accounts in the Late 18th Century

As soon as metallic pipes became used more broadly for the water supply of cities, pipe flow measurements on long conduits were no longer confined to royal gardens. In 1771 Charles Bossut (1730–1814), professor of mathematics at the École du Génie in Mézières, published a treatise on hydrodynamics where he included the results of his measurements from conduits used for the water supply of Mézières (Bossut, 1771, Chap. VI). By and large he agreed with Couplet's findings, so that he allowed himself to proceed one step further to "form a general and sufficient practical idea of the loss of speed that water makes in pipes, either straight or curvilinear."[5] Instead of a formula for the friction he provided tables for the presentation of his results from which a user might choose the case closest to the one he wants to treat.

As a member of the Paris academy, Bossut's expert knowledge was requested also in other hydraulic projects. In 1776, for example, he witnessed water raising trials with the famous machine de Marly. This giant machine pumped water via intermediate reservoirs up to an aqueduct 162 m above the level of the Seine from where it was conducted to Versailles. The plan was to dispose of the intermediate reservoirs and pump the water in a single step up to the aqueduct. However, the pipes used for this purpose proved inadequate. These are matters, Bossut excused the impotence of his expertise in this affair, "whose knowledge is based on practical trial and error and which can only be imperfectly appreciated by theory and reasoning."[6]

Bossut's pipe flow measurements called for more extensive investigations about the friction in long conduits. "I devoured the part of Abbé Bossut's Hydrodynamics, which deals with the movement of water," Pierre du Buat (1734–1809) introduced his *Principes d'Hydraulique, Vérifiés par un grand nombre d'Expériences faites par ordre du Gouvernement*. He regarded the pipe flow data as "the key to hydraulics." Du Buat reasoned that the flow on an inclined river bed reached a uniform speed only because the accelerating force of gravity was balanced by friction, and

> that the movement of water in a pipe had a great analogy with the uniform course of a river bed, since in both cases gravity was the motor, and the resistance of the bed the moderator. I

[4] "... sur la dépense des tuyaux de conduite qui amènent l'eau dans les réservoirs de Versailles, qui est peutêtre le seul endroit du monde où l'on trouve tout ce que l'on peut désirer, pour faire des expériences de la nature de celles dont nous parlons." (Belidor, 1737, p. 273).

[5] "En combinant nos expériences avec celles de M. Couplet, on se formera une idée générale et suffisante dans la pratique, de la perte de vitesse que l'eau fait dans les tuyaux de conduite, soit rectilignes, soit curvilignes." (Bossut, 1771, p. 159).

[6] "... dont la connaissance tient à un tâtonnement pratique et qui ne peuvent être qu'imparfaitement appréciés par la théorie et le raisonnement". Quoted in Brandstetter (2006, p. 186). See also (Barbet, 1907, pp. 141–142).

therefore made use, in order to make a formula of uniform movement, of the experiments of Abbé Bossut on the pipes, and even of those he had made on dummy canals, although one datum was missing, which was the depth of the current. This is how I composed the work that I gave to the public, under the title of Principles of Hydraulics, in the year 1779.[7]

As indicated by the title of du Buat's treatise, his investigations were commissioned by the government, more precisely by the war ministry which also placed at his disposal two officers from the Royal Engineering Corps for performing the experiments. Unlike Bossut's pipes of conduit which had only a "mediocre" slope, du Buat's experiments comprised very small and very large inclinations. The slope was calculated by dividing the length of the pipe by the height of the reservoir from which the water was discharged through the pipes. Furthermore, du Buat and his officers experimented with pipes bent in different ways, comparing the results with those of straight pipes of the same diameter and length. In his treatise, du Buat displayed the results of 56 experiments in tables with five entries for each measurement: the length of the pipe, the height of the reservoir, the slope, the experimentally measured velocity (derived from the discharge divided by the cross section of the pipe), and the velocity calculated from a formula that combined theoretical speculation with empirically adjusted coefficients (Buat, 1786, I, p. xxv and pp. 71–74).

1.3 Early Pipe Flow Formulae

Bossut's and Du Buat's accounts on pipe flow marked the beginning of an era in the history of fluid mechanics in which experiments challenged pre-conceived theoretical concepts about the nature of fluid resistance (Darrigol, 2005, pp. 221–222). Pioneers of ideal flow theory in the 18th century, such as Johann (1667–1748) and Daniel Bernoulli (1700–1782) (father and son) or Leonhard Euler (1707–1783), contributed little to elucidate pipe flow friction. Euler's effort in this regard was largely ignored—and would have been doomed to failure if it had been scrutinized by hydraulic engineers (Bistafa, 2015). Both Bossut's *Traité élémentaire d'hydrodynamique* and du Buat's *Principes d'Hydraulique*, however, were widely appreciated as valuable contributions. They encouraged further investigations on pipe flow and were translated into German. "This book certainly is one of the most useful and best ones that has appeared since some time in hydraulics," Reinhard Woltmann (1757–1837), a German pioneer in hydraulic engineering, praised du Buat's treatise. "Theory and experience go together step by step, and it testifies on almost every page to the skill, diligence and love of truth of the author." The "almost" alluded to du Buat's specu-

[7] "que le mouvement de l'eau dans un tuyau de conduite avait une grande analogie avec le cours uniforme d'un lit de riviere, puisque de part et d'autre la pesanteur était le moteur, et la résistance du lit le modérateur. Je me servis donc, pour faire une formule du mouvement uniforme, des expériences de M. l'abbé Bossut sur les tuyaux de conduite, et même de celles qu'il avait faites sur des canaux factices, quoiqu'il y manquât une donnée, qui était la profondeur du courant. C'est ainsi que je composai l'ouvrage que j'ai donné au public, sous le titre de Principes d'Hydraulique, en l'année 1779." (Buat, 1786, I, pp. xvi–xx).

lations on the nature of friction. "I confess that I don't like the author's imagination on the resistance due to friction," Woltmann qualified his praise. He missed a more detailed consideration of the movement of the water particles close to the wall where, in his view, "they become hindered by collisions, deviated into other directions and into an eddying motion whereby the entire flow is decelerated." Furthermore, Woltmann was unimpressed by du Buat's formula with which the experimental data were fitted. By means of arbitrary quantities "Mr. Buat has modified his formula until it fitted precisely all trials."[8]

Another German expert on hydraulics, Johann Albert Eytelwein (1764–1848), who had started his career in the Prussian administration of constructions and co-founded in 1799 the Berlin Building Academy (Bauakademie), also paid tribute to du Buat in a study "On the motion of water in conduit pipes".[9] He praised Couplet's, Bossut's and du Buat's pipe flow measurements and the effort to arrive at a "general expression" for these results, but he, too, did not like du Buat's formula—even though for other reasons than Woltmann. He regarded the formula as largely useless "because of its complicated form".[10] Eytelwein derived from the same experiments the following formula:

$$u = 6,42\sqrt{\frac{50dh}{l + 50d}},$$

where u is the mean flow velocity in the conduit of pipes, d and l the diameter and length of the conduit, and h the pressure head, i.e. the height between the reservoir and the lower end of the conduit of pipes; all length units in "rheinisch" feet. The discharge M is given by the product of the cross section of the pipe times the velocity

$$M = \frac{1}{4}\pi d^2 u = 5.04 d^2 \sqrt{\frac{50dh}{l + 50d}}.$$

Eytelwein's pipe flow formula did not fit the data as well as du Buat's formula—but had the virtue of being useful for determining the dimensions of pipes for given purposes. It could be used to solve engineering problems such as: "How large has to

[8] "Dieses Buch ist gewiss eines der nützlichsten und besten, die seit einiger Zeit in der Hydraulik erschienen sind. Theorie und Erfahrung gehen in demselben Schritt für Schritt beisammen; und es zeugt von der Geschicklichkeit, dem Fleiße und der Wahrheitsliebe des H. Verf. fast auf allen Seiten. [...] Ich gestehe, dass mir des H. Verf. Vorstellungsart über den Widerstand wegen des Reibens nicht gefällt [...] Dagegen stoßen die Teilchen an, werden aufgehalten, in andere Richtungen und mancherlei wirbelnde Bewegung gebracht, wodurch der ganze Strom verzögert wird [...] hat Hr. Buat seine Formel so lange abgeändert bis sie auf alle Versuche genau passte." (Woltmann, 1791, pp. 142, 152, 162). On Woltmann's career see (Ruehlmann, 1885, pp. 280–283).

[9] "Von der Bewegung des Wassers in Röhrenleitungen". (Eytelwein, 1801, 9. Kapitel). On Eytelwein's career see (Ruehlmann, 1885, pp. 284–289).

[10] "Herr du Buat hatte das groß e Verdienst, zuerst einen allgemeinen Ausdruck mitgetheilt zu haben, welcher mit den bekannten Erfahrungen übereinstimmt, und der bloß den Fehler hat, dass er wegen seiner verwickelten Form, nur mit vieler Weitläufigkeit Anwendung findet." (Eytelwein, 1801, p. 216). Du Buat's formula is discussed in (Rouse and Ince, 1957, pp. 131–132).

be the diameter of a straight conduit pipe with a length of 100 ft at a pressure head of 5 ft in order to discharge a half cubic feet per second?"[11]

In 1804 Gaspard de Prony (1755–1839), director of the École des Ponts et Chaussées and perhaps the most prominent representative of hydraulic engineering in France, made a major step forward by suggesting a formula which combined pipe flow and open channel flow. He shared the view—introduced by Charles Coulomb (1736–1806) a few years earlier in the wake of his famous torsion balance experiments— that there are two forces involved in fluid friction: one due to the mutual interaction of the molecules of the fluid, and one due to the collision of the fluid molecules with surface irregularities at the bounding wall. The former resulted in a friction proportional to the flow velocity, the latter implied an inertial retardation proportional to the square of the velocity (Darrigol, 2005, pp. 104–106). Prony, therefore, attempted to account for the experimental data by making the ansatz

$$\frac{gdh}{4l} = au + bu^2$$

where g is the gravitational acceleration and a, b are constants to be determined from measurements. Like Eytelwein, Prony relied on the pipe flow experiments of Couplet, Bossut and Du Buat. Furthermore, he employed the newly introduced metric system. Altogether, the set of data comprised the results of 51 experiments with conduits of tin and cast iron pipes that differed considerably in diameter (from 2.7 to 49 cm) and length (from 9.7 to 2280 m). For the determination of a and b Prony adopted a method developed by Pierre-Simon Laplace (1749–1827) in celestial mechanics for matching observational data to mathematical expressions. As a result, Prony's *Recherches Physico-Mathematiques sur la Théorie des Eaux Courantes* represented empirical hydraulic knowledge combined with sophisticated mathematical elaboration like no other contemporary account (de Prony, 1804).

Prony's formula dominated hydraulic engineering for several decades. The two terms proportional to u and u^2 seem to foreshadow the distinction of laminar and turbulent flow, but this is a retrospective interpretation. The notion of viscosity as introduced by Claude-Louis Navier (1785–1836) in the 1820s in the fundamental equations of flow did not enter pipe flow formulae before the second half of the 19th century (Darrigol, 2005, Chap. 3).

1.4 Air Flow in Long Conduits of Pipes

While Prony's formula made the pipe flow of water accessible to hydraulic engineering, the flow of air in conduits of pipes remained mysterious—although the

[11] "Wie groß wird man den Durchmesser einer graden 100 Fuß langen Röhrenleitung bei einer Druckhöhe von 5 Fuß annehmen müssen, damit solche in jeder Sekunde einen halben Kubikfuß Wasser liefert?" Inserting $M = \frac{1}{2}, h = 5, l = 100$ yields an equation for d with the result $d = 0,341$ ft (Eytelwein, 1801, p. 223).

latter was of comparable practical relevance in the industrial revolution. In order to provide mine air or operate blast furnaces, air had to be pumped through conduits of pipes with different diameter and length. In 1797, Joseph Baader (1763–1835), an engineer at the Bavarian court who had spent his formative years in England and Scotland, relayed the following experience of the English "iron master" John Wilkinson (1728–1808) about the procurement of air for a blast furnace:

> He [Wilkinson] intended to use the steep slope of a stream for the operation of a blast furnace 5000 feet away. He built a large overshot water wheel with a cylindrical blowing engine and installed an air conduit with 12 inch wide pipes of cast iron straight towards the furnace. When the machinery was completed and the water-wheel set in motion for the first time, compressed air escaped to the astonishment of all present through the most minute openings and through the valve of the machine itself, while at the other end of the conduit of pipes there was not the slightest motion of air as was checked by the flame of a candle! Then one carefully sealed all openings and fastened the valve so that it would withstand the pressure of the compressed air. When the water wheel was set in motion again, it turned around slower and slower until it stood still entirely. Although the air inside the blowing engine was now compressed to the extent that its elasticity balanced the whole applied force there was not the slightest current of air at the other end of the conduit. Naturally suspicion was raised that the conduit was accidentally blocked at some point. In order to check this hypothesis, a cat was put inside the entrance of the pipe which was then closed; but the cat happily walked out from the other end and therefore had met no obstruction! [12]

According to Baader's report Wilkinson further investigated the mysterious blockade of air flow by drilling holes at a distance of 30 ft from another in the pipes. Thus Wilkinson observed how the escape of air through these holes became weaker and weaker further away from the blowing engine, until at a distance of 600 ft only a faint breath of air was noticed. Baader found it regrettable that the pressure at the drill holes was not measured by barometer tubes. Thus one had missed the opportunity to uncover the law according to which the resistance of a mass of moving air in a long

[12] "Er gerieth auf den Einfall, einen Bach mit einem starken Gefälle zur Betreibung eines Hohofens zu benützen, der 5000 Fuß, (ohngefähr eine englische Meile) von der Stelle entfernt war. In dieser Absicht baute er ein großes oberschlächtiges Rad mit einer vollständigen Zylindermnschine, und führte eine Windleitung von 12 Zoll weiten gegossenen eisernen Röhren von der Maschine gerade nach dem Ofen. Als nun die ganze Anlage vollendet war, und man das erstemal Wasser aufs Rad schlug, zeigte sichs zum großen Erstaunen aller Gegenwärtigen, dass die zusammengepreßte Luft durch die kleinsten Öffnungen und Fugen, vorzüglich aber durch ein mit Gewicht beschwertes Ventil (Wastevalve) an der Maschine selbst entwischte, indes aus der Öffnung am entfernten Ende der Röhrenleitung durch ein vorgehaltenes Licht nicht einmal die geringste Bewegung zu bemerken war!—Man verstopfte hierauf alle Fugen auf das sorgfältigste, und beschwerte das Ventil nach und nach mit soviel Gewicht, dass die verdichtete Luft solches gar nicht mehr zu heben vermögend war, und das Rad, bey vollem Aufschlagwasser, sich immer langsamer und langsamer bewegte, bis es endlich ganz stille stand. Allein obwohl nunmehr die Luft in der Maschine offenbar auf einen so hohen Grad verdichtet war, dass ihre Elastizität der ganzen vorhandenen Kraft das Gleichgewicht hielt, es war doch an dem entfernten Ende der Windleitung noch nicht der schwächste Luftzug zu spüren. Natürlicherweise entstand jetzt der Verdacht, dass die Röhrenstrecke an irgend einer Stelle durch einen Zufall verstopft wäre, und um diese Hypothese zu prüfen, steckte man in die Mündung der Windleitung bey der Maschine eine lebende Katze, welche, nachdem ihr der Rückweg verschlossen ward, nach einiger Zeit an dem andern offenen Ende (von welchem das enge Blaserohr abgenommen war) glücklich herauskam, folglich die ganze Röhrenleitung ohne Widerstand durchlaufen hatte!" (Baader, 1797, pp. 165–166).

conduit of pipes increases with the length of the conduit (Baader, 1797, p. 167). He reported about Wilkinson's experience again at another opportunity (Baader, 1805, pp. IX–XI), but measurements on the flow of air in long pipes were performed only several years later. In July 1811 the *Journal de Physique* reported about experiments of air flow in a 447.5 m long iron pipe with a diameter of 25 cm. In this case no blockade was observed, although the air flow was caused by only a tiny fraction of the power of Wilkinson's blowing engine (Lehot et al., 1811).

In 1819 the friction of the air flow became the subject of more systematic measurements at a hospital in Paris where the new technology of gaslight had been introduced two years before. Pierre Simon Girard (1765–1836), a former collaborator of Prony and an expert on hydraulics, used the opportunity to measure the flow of gas driven by a gasometer through a 623 m long conduit of pipes with a diameter of 81 mm. By comparing the flow of coal gas with that of air, both driven by the same pressure from the same gasometer, Girard ascertained that gaseous fluids move through the pipes more or less independent of their compressibility ("indépendamment du plus ou moins d'élasticité dont ils peuvent être doués"). From this observation he concluded that even incompressible fluids should obey the same law of friction:

> Therefore, these different phenomena of gas motion are exactly the same as those of incompressible liquids, from which we may conclude that the same formulae must serve to calculate the movement of both.[13]

Accordingly Girard applied Prony's formula derived from experiments on the flow of water to the flow of gases:

$$\frac{gdhp}{4p'l} = au + bu^2$$

where p and p' are the specific weights of water and and gas, respectively, h the height of the water column that drives the gas through the conduit of pipes with length l and diameter d. Different gases would be matched by this formula with different coefficients a and b, but for very long pipes the term proportional to u seemed negligible, and then the coefficients b differed only little.

The experiments with the pipes of the Paris gaslight did not involve very large flow speeds like those encountered in the pipes of blowing engines for blast furnaces or mine air. In order to fill this gap, Jean-François d'Aubuisson de Voisins (1769–1841), chief engineer of the royal mining corps, performed experiments on the flow in large pipes used to provide air for an iron mine in the department of l'Ariége. 20 m long pipes made of tin with a diameter of 10 cm were joined together to a 387 m long conduit. The pressure in the pipes was measured at drill-holes 40 m apart by manometers (double-bent glass tubes filled with mercury). In the course of several years, d'Aubuisson concluded from a great number of measurements that indeed "the resistance is proportional to the square of the velocity", as Girard already had

[13] "Or, ces divers phénomènes du mouvement des gaz sont exactement les mêmes que ceux du mouvement linéaire des liquides incompressibles, d'où il est permis de conclure que les mêmes formules doivent servir à calculer le mouvement des uns et des autres." (Girard, 1821, p. 144).

assumed for larger pipes, so that only a single coefficient remained to be determined from the experiments. In the same vein his results corroborated Eytelwein's formula which, as d'Aubuisson noticed, comes out as the same as that of Prony as long as the length of the conduit is very large compared to the diameter (Aubuisson, 1828, p. 444).

1.5 The Darcy-Weisbach Equation

Besides the experiments on air flow in long pipes, d'Aubuisson was involved in the design of a new water supply system of Toulouse.[14] When he reviewed the contemporary knowledge on hydraulics as far as it was "useful for engineers" he divided his account in one part on hydraulics proper ("L' hydraulique proprement dite") and another part on air measurements ("L' aerometrie") (d'Aubuisson de Voisins, 1834). Hence he assigned pipe flow of water and air to separate chapters, but he argued that the same forces were responsible for the flow retardation in both cases:

> Since the resistance comes from the action of the walls, it will be proportional to their extent, i.e., to the length of the pipe and the perimeter of the cross-section [...] On the other hand, the larger the cross-section, the more the resistance of the walls will be distributed over a greater number of molecules and therefore will affect each of them and the total mass less, and will therefore be inversely related to this number, hence inversely proportional to the cross-section.[15]

Prony's and Eytelwein's formulae could be regarded as the mathematical expression of this view. They provided the theoretical basis from which in most practical cases the required quantities of a conduit of pipes could be calculated—for water as well as for gases. Yet the friction of fluids in pipe flow remained a major challenge to bring theory in line with practice. In 1839, Gotthilf Hagen (1797–1884), a professor of hydraulic engineering at the Berlin Bauakademie, raised "some doubts" about Prony's and Eytelwein's formulae. The influence of temperature, for example, "is completely ignored in these hydraulic formulae." (Hagen, 1839, p. 423) He performed a series of experiments on the flow of water at different temperatures through a narrow, mean and larger pipe with diameters of 2.6 mm, 4 mm and 5.9 mm, respectively. Like others before him, Hagen noticed a dependency of the pipe friction on temperature, but he postponed the detailed study of this influence for later (Hagen, 1855). For the time being he derived from his experiments a formula which would later be called after him and Jean Léonard Marie Poiseuille (1797–1869), the

[14] Notice Nécrologique sur M. d'Aubuisson de Voisins in: Annales des Mines, 4eme série, tome 11, 1847, pp. 667, available at http://www.annales.org/archives/x/aubuisson.html#annales.

[15] "Puisque la résistance provient de l'action des parois, elle sera proportionnelle à leur étendue, c'est à-dire, à la longueur de la conduite et au périmètre de la section [...]. D'un autre côté, plus la section sera grande, et plus la résistance des parois se distribuera sur un plus grand nombre de molécules; par conséquent elle affectera moins chacune d'elles et la masse totale : elle sera donc en raison inverse de ce nombre, lequel dépend de la grandeur de la section." (Aubuisson, 1834, p. 171).

law of laminar pipe flow.[16] Hagen's textbook reputation is confined to the "Hagen-Poiseuille law" only, but for his contemporaries he was an authority for the entire field of hydraulic engineering. He authored a voluminous *Handbuch der Wasserbaukunst* and served as an expert adviser for water works and other building projects in Prussia. In his pipe flow studies he did not content himself with establishing a formula for laminar flow, but also observed the transition to turbulent flow:

> When I let the water flow freely in the air, for smaller pressure heads the jet had a permanent shape, and near the pipe it looked like a solid piece of glass; but as soon as the velocity, by stronger pressure, exceeded the given limit, the jet started to fluctuate and the efflux was no longer uniform and occurred by pulses.[17]

When Hagen investigated a few years later the influence of temperature more closely he added saw dust to the water in order to visualize these "two kinds of motion":

> I observed that for small pressures the saw dust propagated only in the direction of the tube, whereas for strong pressures it shot from one side to another and often assumed an eddying motion.[18]

Thus Hagen's pipe flow investigations came close to discoveries that are ascribed to Osborne Reynolds thirty years later (Rouse and Ince, 1957, p. 160; Darrigol, 2005, p. 256).

Another German hydraulic engineer who paid great efforts to improve Prony's and Eytelwein's pipe flow formulae was Julius Weisbach (1806–1871). In 1845 he published the first of three volumes of a textbook on mechanics, explicitly for the use of engineers, with an extensive section on the "dynamics of fluid bodies". With regard to pipe flow he added to the 51 measurements of Couplet, Bossut and du Buat one experiment of an hydraulic engineer at Grenoble (Émile Gueymard, 1788–1870) and 11 of his own by which the range of flow velocities in the pipes was considerably increased towards higher speeds. Weisbach arrived at the formula

$$h = f\frac{l}{d}\frac{u^2}{2g}$$

[16] For a discussion of Hagen's and Poiseuille's experiments see (Sutera and Skalak, 1993; Darrigol, 2005, Sect. 3.7).

[17] "Ließ ich nämlich das Wasser frei in die Luft ausströmen, so bildete der Strahl bei kleineren Druckhöhen eine unveränderte Form, und er hatte in der Nähe der Röhre das Ansehen eines festen Glasstabes; sobald aber bei stärkerem Drucke die Geschwindigkeit die bezeichnete Grenze überstieg, so fing er an zu schwanken und der Ausfluss geschah nicht mehr gleichförmig, sondern stoßweise." (Hagen, 1839, p. 423), translated in (Darrigol, 2005, p. 244).

[18] "Indem ich durch dieselbe Röhre zugleich mit dem Wasser auch Sägespähne hindurchtreiben ließ, so bemerkte ich, dass dieselben bei geringem Drucke nur in der Richtung der Röhre fortschritten, bei starkem Drucke dagegen, von der einen Seite zur andern geschleudert wurden, und oft in wirbelnde Bewegung geriethen." (Hagen, 1855, p. 81), translated in (Rouse and Ince, 1957, p. 159; Darrigol, 2005, p. 245).

with the friction factor $f = a + \frac{b}{\sqrt{u}}$; he determined a and b by the method of least squares. This formula, Weisbach concluded, fits the measurements "much more precisely" than the formulae by Prony and Eytelwein (Weisbach, 1845, p. 433).

New pipe flow measurements were regarded imperative also in the course of other hydraulic projects such as for the water supply at Dijon implemented in 1840 by Henry Darcy (1803–1858), a young engineer from the Corps des Ponts et Chaussées. *Les Fontaines Publiques de la Ville de Dijon*, as Darcy titled the treatise about his accomplishment (Darcy, 1856), represented one of the most modern water supply systems by this time. "The execution of the work on the Dijon water supply gave me the opportunity to recognize the need to engage in the *Recherches* I am publishing today", Henry Darcy introduced a subsequent treatise.[19] Darcy's memoirs are celebrated as landmarks in the history of hydraulics and secured him a place in the hall of fame of this discipline (Rouse and Ince, 1957; Freeze, 1994; Brown, 2002a; Gisonni, 2003; Hager, 1994; Bobeck, 2006; Simmons, 2008, pp. 169–172).

The opportunity to perform pipe flow experiments on an heretofore unprecedented scale came when Darcy was charged in 1850 with the supervision of the Paris water supply. As he recalled in his treatise about the Dijon fountains:

> My original thought was to seek the means to verify the exactness of the laws that are generally accepted today. However, it did not take me long to recognize that I did not have in hand the elements necessary to resolve such a basic question. I had to wait to take up this question until circumstances permitted. I was able to take up these earlier thoughts when I was called to Paris as Director of Municipal [Water] Service. I then had at my disposal all the equipment necessary to carry out the experiments that I had thought about for so long.[20]

In his experiments, Darcy wished to clarify in particular the influence of the pipe wall roughness and diameter on the flow resistance. He measured the flow profile across the cross section of a pipe by an improved Pitot tube (Brown, 2003) and compared the flow of water through 100 m long conduits of pipes of different materials and diameters between 1 cm and 50 cm. Furthermore, he employed large hydraulic heads in order to achieve large flow velocities. A good deal of his data such as those from old and rough pipes could not be reconciled with Prony's formula. Darcy arrived at a formula for the head loss which contained a linear and a quadratic term in the velocity, like that of Prony, but with coefficients that displayed a marked dependency on the pipe diameter. Furthermore, Darcy noticed that beyond a certain velocity the linear term could be neglected.[21] On the other hand, for pipes with very small diam-

---

[19] "L'exécution des travaux relatifs à la fourniture d'eau de Dijon m'a donné lieu de reconnaître la nécessité de me livrer aux *Recherches* que je publie aujourd'hui." (Darcy, 1857, p. V).

[20] "Ma première pensée avait été de chercher dans ces expériences les moyens de vérifier l'exactitude des lois admises jusqu'à ce jour; mais je n'ai pas tardé à reconnaître que je n'avais pas sous la main les éléments nécessaires pour résoudre une si grave question, et j'ai attendu que les circonstances me permissent de songer à la reprendre. Je donnai suite à ma pensée première lorsque je fus appelé à Paris, comme directeur du service municipal. J'avais alors à ma disposition tous les appareils indispensables pour procéder aux expériences auxquelles je songeais depuis si longtemps." (Darcy, 1856, p. 376). English translation in (Bobeck, 2006, p. 1003).

[21] "... qu'à partir d'une certaine vitesse la valeur du débit ne semble point affectée par la suppression du premier terme." (Darcy, 1857, p. 120).

eter and flow velocity, the quadratic term could be neglected. "M. Poiseuille and I arrived at this expression by totally different experiments and circumstances", Darcy remarked about this case which we now call laminar flow. Darcy contented himself with the remark that "it seems that two laws appear in the phenomenon produced by the flow of water in the pipes and that these laws, at least in the diameters that I have used, merge at velocities of 9–10 cm [per second]."[22]

For other pipes, in particular those with rough walls, Darcy's formula reduced to an expression proportional to the square of the velocity where the resistance could be expressed by a friction factor—similar to the one published a few years earlier by Weisbach (whose work seems to have been unknown to Darcy). Unlike Weisbach's friction factor (see above), however, Darcy's factor showed a functional dependence on the pipe diameter: $f = a + \frac{b}{d}$ (Darcy, 1857, p. 117). Like Weisbach, Darcy determined the coefficients (a, b) by the method of least squares from the experimental measurements. Other features (pipe roughness, viscosity of the fluid) were not made explicit.

It was left to another hydraulic engineer in the U.S.A., John Thomas Fanning (1837–1911), to adjust Weisbach's expression to Darcy's better data and to establish what became known as "Darcy-Weisbach equation" (Brown, 2002b). The modern understanding of this law had to await the similarity theory as applied by Heinrich Blasius (1883–1970) to empirical pipe flow data (see below). In the same vein, Darcy has been praised for "the first accurate measurements of turbulent pipe velocity distributions" and for providing "the very first evidence of the existence of the fluid boundary layer" (Simmons, 2008, p. 1028). Both ascriptions need some qualification: Darcy used erroneous assumptions to derive from his measurements the velocity profiles (Gisonni, 2003, p. 34); although he located the origin for the wall friction in "the small liquid band in contact with the walls" where "the fluid elements are agitated in vortical motions" ("la petite couche liquide en contact avec les parois... les éléments de cette couche sont animés de mouvements giratoires") he made no attempt to determine the features of this layer (Darcy, 1857, p. 10).

Darcy's pipe flow measurements became the prototype to which other hydraulic engineers referred in the design of water supply systems elsewhere. After Darcy's death in 1858, his assistent Henri Bazin (1829–1917) extended these investigations to the study of open channel flow where the walls seemed to exert a similar flow retardation as in closed pipes.[23] The results of Darcy's and Bazin's experiments launched a new quest for a general flow formula (Hager, 1994).

In 1874, the Union of German Architectural and Engineering Associations solicited a review on "measurements concerning the pressure head loss in the flow

---

[22] "... nous sommes parvenus, M. Poiseuille et moi, à cette expression, au moyen d'expériences faites dans des circonstances tout à fait différentes.... il semble que deux lois apparaissent dans le phénomène produit par l'écoulement de l'eau dans les tuyaux de conduite et que ces lois, au moins dans les diamètres que j'ai employés, viennent se souder vers les vitesses de 9 à 10 centimètres." (Darcy, 1857, pp. 213–215).

[23] "L'influence capitale qu'exerce la nature de la paroi était incontestable en principe après les expériences faites sur les tuyaux de conduite par M. Darcy; elle se trouve donc démontrée d'une manière encore plus frappante pour les canaux découverts." (Darcy and Bazin, 1865, p. 79).

of water through conduits of cast iron pipes of various pipe width, flow velocity and under the influence of successive modification of the inner surfaces". The motivation for this review came from the uncertainty "which of the numerous formulae is to be taken as the basis for any new installations as being closest to the truth." As a result of this request new pipe flow measurements in several German cities (Danzig, Bonn, Wiesbaden, Hamburg) were performed and analysed with regard to various formulae. By and large, the validity of Darcy's formula was found to agree best for new pipes, even for pipe diameters as large as 122 cm, but no formula could account for older pipes with rough inner surfaces.[24]

1.6 The Riddle of Two Different Manners of Flow

By the 1860s the experiments of hydraulic engineers like Hagen, Weisbach, Darcy and others had abundantly made clear that the flow in very narrow pipes was fundamentally different from that in large pipes. The former had been analysed by Hagen and Poiseuille whose law could be derived from the Navier-Stokes equations. In the contemporary parlance of the year 1860, capillary flow represented "the only form of experiment for which the complete theory could be derived from those equations of motions", whereas the experimental results of pipe flow in larger pipes could only be used to derive "practical approximation formulae but not for comparison with a mathematical theory".[25]

The failure of fundamental hydrodynamic theory to cope with this problem became notorious. In 1872, Adhémar Barré de Saint-Venant (1797-1886), the outstanding expert on theoretical mechanics among the French academicians, regarded this problem "a true enigma".[26] His pupil Joseph Boussinesq (1842–1929) echoed this view: "Fluids move in two different manners, according to whether they flow in very narrow tubes or in spaces having sections comparable to those of pipes and open channels." Thus Boussinesq introduced a treatise on the flow of water. In the latter case one had to take into account "the eddying agitation" that extends from the walls into the interior of the fluid and propagates very fast or even abruptly from point to

[24] "Welche Erfahrungen, bezw. Messungen liegen vor über den Druckhöhenverlust beim Durchfluss von Wasser durch gusseiserne Rohrleitungen für verschiedene Rohrweiten, verschiedene Geschwindigkeiten und unter dem Einfluss der successiven Oberflächenveränderung im Innern? ... welche der zahlreichen Formeln, als der Wahrheit am nächsten kommend, bei etwaigen Neuanlagen zu Grunde zu legen sei" (Iben, 1880, p. 1).

[25] "Indessen ist bisher nur eine einzige Form des Experimentes gefunden worden, für welche die vollständige Theorie aus jenen Bewegungsgleichungen hergeleitet werden kann, es ist das die Bewegung einer Flüssigkeit in sehr engen und sehr langen Röhren. [...] so dass man die unter solchen Bedingungen gewonnenen Ergebnisse zwar zur Ableitung praktischer Approximationsformeln, aber nicht zur Vergleichung mit der mathematischen Theorie gebrauchen kann." (Helmholtz and Piotrowski, 1860, pp. 607–608). See also (Darrigol, 2005, Sect. 3.7).

[26] "une véritable enigme". Rapport sur un mémoire de M. Boussinesq, présenté le 28 Octobre 1872, in (Boussinesq, 1877, p. III).

point.[27] Boussinesq combined Saint-Venant's theoretical views with the engineering knowledge about pipe and channel flow as described in the treatise of Darcy and Bazin. From "the eddying agitation" he derived the concept of an eddy viscosity which would play a fundamental role in theories of fully developed turbulence in the 20th century.

Another effort to bring the two different manners of flow under a common theoretical roof was made a few years later by Osborne Reynolds (1842–1912), an engineering professor at Manchester with a tendency towards research problems that have "both a practical and a philosophical aspect". This was how he described his objective for "An Experimental Investigation of the Circumstances which determine whether the Motion of Water shall be Direct or Sinuous, and of the Law of Resistance in Parallel Channels" (Reynolds, 1883)—a landmark paper in the history of turbulence (Fig. 1.2). Reynolds has often been credited for achievements which were made by others. Hagen visualized the transition to turbulence many years before Reynolds (see above). Another part of the "Reynolds legend" (Darrigol, 2005, p. 255) concerns the rise of similarity theory. Reynolds did not claim priority for the study of the scaling invariance of the Navier-Stokes equation. "I had no intention whatever of laying down the conditions of dynamical similarity," Reynolds wrote in 1883 in a letter to Gabriel Stokes (1819–1903) [quoted in (Darrigol, 2005, p. 256)]. The late rise of the "Reynolds number" as a key concept in fluid mechanics—it received this name only in 1908 (Rott, 1990)—further hints at the haphazard use of similarity theory.

Reynolds's achievements have been analysed in great detail and need no further review (Launder and Jackson, 2011; Darrigol, 2005, Sect. 6.5), but it seems appropriate to focus on some of his findings once more from the perspective of pipe flow. What we now call "Reynolds number" emerged in bits and pieces with the attempt to determine the critical velocity ("at which eddies come in") for different pipes, viscosities and flow velocities. Reynolds displayed the loss of pressure in a pipe in a double-logarithmic plot ("logarithmic homologue") as a function of the flow velocity in order to expose the common features of different pipe flows. The data collapsed to straight lines with different "Log Slopes of Pressure" that intersected at the critical velocity (Fig. 1.3). Reynolds was surprised that he could even accommodate Darcy's data in his analysis although they referred to pipes of quite different dimensions. "I could not possibly expect any close agreement with my results; still, as experiments on pipes of such large diameters are not likely to be repeated, at any rate with anything like the same care and success, they offered the only chance of proving that my law was general" (Reynolds, 1883, p. 977).

[27] "Les fluides se meuvent, de deux manières différentes, suivant, qu'ils coulent dans des tubes très-étroits ou dans des espaces ayant des sections comparables à celles des tuyaux de conduite ou des canaux découverts. [...] Au reste, cette production d'une agitation tourbillonnàire au sein de toute masse fluide qui s'écoule à travers des sections d'une certaine étendue [...] Il faut donc... regarder les vitesses vraies, à l'intérieur d'un fluide qui s'écoule, comme rapidement ou même brusquement variables d'un point à l'autre, capables, en un mot, de produire des frottements d'un tout autre ordre de grandeur que dans le cas de mouvements continus... (Boussinesq, 1877, pp. 1, 6–7). See also (Rouse and Ince, 1957, p. 203; Darrigol, 2005, p. 235).

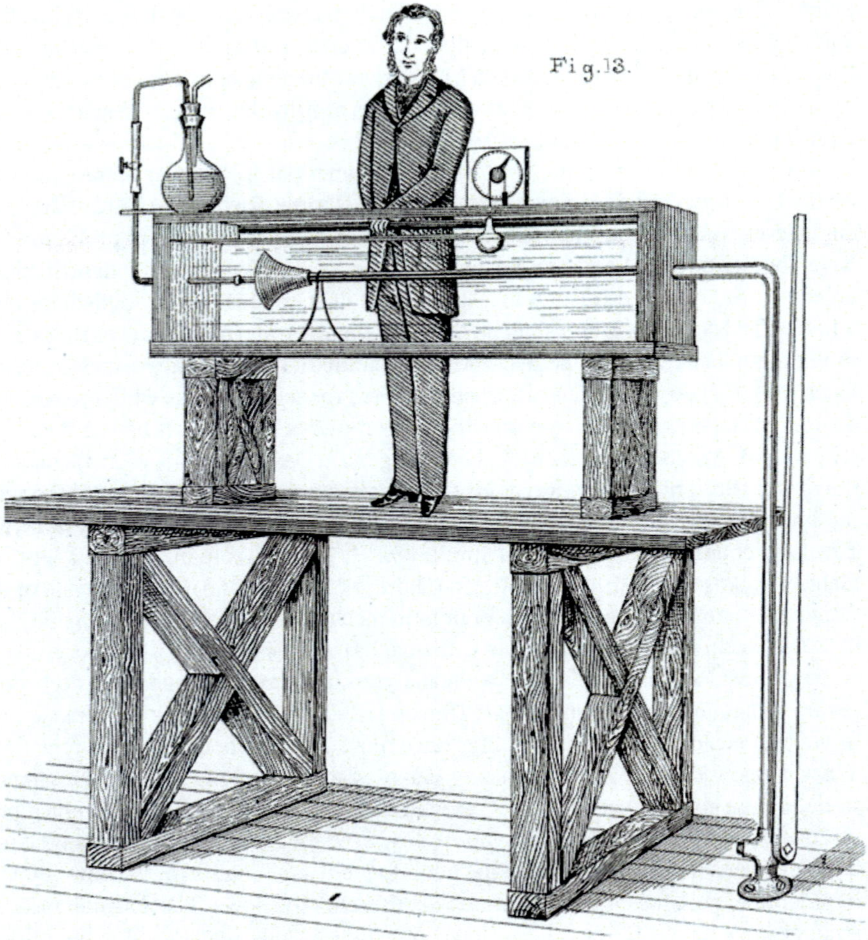

Fig. 1.2 The iconic display of the experiment with which Reynolds demonstrated the transition from laminar to turbulent pipe flow (Reynolds, 1883, plate 73)

In a subsequent paper published twelve years later Reynolds expressed the critical velocity as $K\mu/D\rho$, where D is the pipe diameter, μ the viscosity and ρ the density of the fluid, and K "a numerical constant, the value of which according to my experiments, and, as I was able to show, to all the experiments by Poiseuille and Darcy, is for pipes of circular section between 1900 and 2000" (Reynolds, 1895, p. 125). Thus the divide between the two manners of flow, "steady direct" and "sinuous or eddying" in Reynolds's parlance, was for the first time quantified and subjected under a common theoretical roof. However, it took many years until Reynolds's achievement became "a marker for the direction of research in Engineering Fluid Mechanics for the next century" (Jackson and Launder, 2007, p. 20). Contemporary

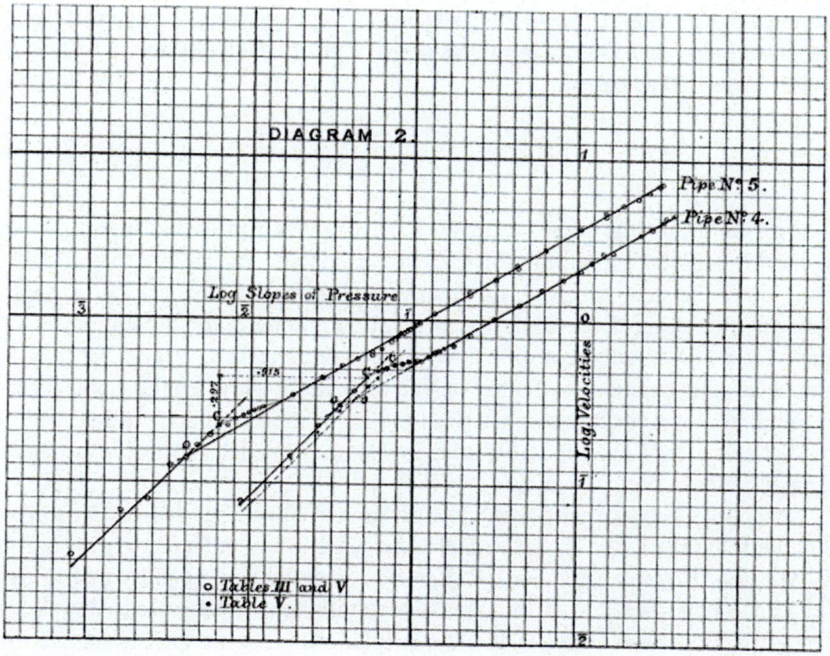

Fig. 1.3 Reynolds's double-logarithmic plot of the loss of pressure versus velocity displays the two manners of flow by different "Log Slopes of Pressure". It yields the critical velocity at the intersection of straight lines (Reynolds, 1883, plate 73)

hydraulic engineers hardly recognized Reynolds' papers from 1883 and 1895 as key contributions to their field. A review on pipe and channel flow from 1897 pointed to the work of Hagen and others who had investigated the two different manners of flow much earlier: "Reynolds has therefore been anticipated five or six times." Furthermore, there were experimental pipe flow data "adverse to Reynolds' theory, which therefore may be rejected" (Knibbs, 1897, pp. 326, 343).

1.7 Blasius' Law for Turbulent Pipe Friction

In 1903, two American engineers, Augustus V. Saph and Ernst W. Schoder (1879–1968), presented the results of pipe flow measurements of heretofore unsurpassed extent and precision (Saph and Schoder, 1903). In hundreds of experiments with different types of pipes of different diameters they measured the resistance of the flow of water at different temperatures (thus modifying viscosity like Hagen and others). They displayed the data as double-logarithmic plots for the head loss per unit length as a function of the flow velocity. The plots for different pipes could be grouped as parallel straight lines with two distinct slopes representing the two manners of pipe flow, as Reynolds had shown twenty years ago. From their diagram they concluded that for smooth pipes above a critical velocity the head loss h per length l was "very nearly"

$$\frac{h}{l} \sim \frac{u^{1.75}}{d^{1.25}}.$$

With hindsight the same result may be uncovered in earlier investigations of hydraulic engineers, but none of these were specific for smooth pipes only and disposed of the same precision for determining the exponents of u and d. Saph and Schoder, therefore, deserve particular mention among hydraulic engineers in the quest for a law of turbulent pipe flow (Steen and Brutsaert, 2017). Yet they—like most hydraulic engineers—did not pursue their study with "a philosophical aspect" in mind, as Reynolds had characterized the motivation for his investigations. Saph and Schoder cited Reynolds only for having demonstrated the change in the character of pipe flow at a critical velocity.

Reynolds's papers found more interest among physicists. It was left to Arnold Sommerfeld (1868–1951), theoretical physicist at Munich University, to introduce in 1908 the "Reynolds number" (Rott, 1990) as the critical quantity in a study that aimed at a determination of the transition to turbulence in the plane flow between two walls moving in opposite directions (plane Couette flow) (Eckert, 2010). By the same time the experimental physicist Wilhelm Wien (1864–1928) at Würzburg University asked a doctoral student to investigate the occurrence of turbulence phenomena more closely in gases instead of water, with a particular interest in the range of the Hagen-Poiseuille law. Wien had published in 1900 a textbook on hydrodynamics where he lamented about the gap between theory and practice in this area: "Here the actual processes are so inconsistent with the theoretical conclusions that engineering developed a special way of treating hydrodynamic tasks, which is usually called hydraulics."[28] The experiments of Wien's doctoral student clearly showed "that the critical value of Reynolds is also decisive for the beginning of turbulent air flow through narrow pipes."[29]

The next step was made by Heinrich Blasius at the Berlin Research Laboratory for Hydraulics and Naval Architecture (Königliche Versuchsanstalt für Wasserbau und Schiffbau). Blasius was prepared for this investigation like few others (Hager, 2003). He had finished his studies at Göttingen University as Ludwig Prandtl's (1875–1953) (Eckert, 2017) first doctoral student with an elaboration of Prandtl's boundary layer concept (see Chap. 5). In 1907/08 he reviewed the contemporary knowledge on turbulence in a seminar (Eckert, 2019a). When he moved in 1908 from the academic

[28] "Hier stimmen die thatsächlichen Vorgänge mit den theoretischen Folgerungen vielfach so ungenügend überein, dass die Technik sich für ihre Zwecke eine besondere Behandlungsweise hydrodynamischer Aufgaben, die meistens den Namen Hydraulik führt, zurechtgemacht hat." (Wien, 1900, p. III).

[29] "Bei Versuchen mit Strömungen der Luft durch enge Röhren hat sich ergeben, dass der kritische Wert von Reynolds auch hier maßgebend ist für den Beginn turbulenter Strömung." (Ruckes, 1908, p. 1020).

environment in Göttingen to the engineering laboratory in Berlin he was eager to demonstrate his expert knowledge in practical applications. He became aware that hydraulics offered many opportunities for similarity considerations. Even after thirty years "the law of Reynolds" had hardly penetrated engineering practice, he noted in the first of several same-titled papers on "The law of similarity in frictional processes". When the motion was determined by frictional and inertial forces like in pipe flow, the law of similarity implied that the friction coefficient was a function of the Reynolds number $\frac{ud}{\nu}$, where ν is the kinematic viscosity of the fluid. Besides the friction in pipe flow the similarity law provided the basis for wind tunnel tests like those in Prandtl's new Göttingen Airship Model Research Institute (Luftschiff-Modellversuchsanstalt) (Eckert, 2017, Sect. 4.1). "There is then no doubt that the introduction of the law in practice is necessary," Blasius concluded this paper.[30]

In November 1911 he presented the same message to a conference on aeronautics in Göttingen—with the focus on the flow of air instead of water. He concluded from recent measurements about the head loss of air flow in pipes that "these experiments carried out with atmospheric air yielded the same coefficient for the same Reynolds number as the series of experiments with water mentioned above." When he was asked in the discussion about the difference between laminar and turbulent flow he presented the Darcy-Weisbach-formula for the head loss

$$h = \lambda \frac{l}{d} \frac{u^2}{2g}$$

with the friction factor for laminar flow

$$\lambda = 64 \left(\frac{ud}{\nu} \right)^{-1}$$

and for turbulent flow

$$\lambda = 0.3164 \left(\frac{ud}{\nu} \right)^{-1/4}$$

"The transition from one law to the other takes place at values of $\frac{ud}{\nu}$ between 2000 and 3000."[31] The "other" law for turbulent pipe friction became known as "Blasius' law".

For the extensive presentation of his results Blasius chose the Association of German Engineers (Verein Deutscher Ingenieure) as his audience (Blasius, 1913). In all of his publications he left no doubt about the importance of experiments because

---

[30] "Es ist danach kein Zweifel, daß die Einführung des Gesetzes in die Praxis notwendig ist." (Blasius, 1911, p. 1177).

[31] "Diese mit atmosphärischer Luft angestellten Versuche ergaben für gleiche Reynoldssche Zahl denselben Koeffizienten, wie die oben erwähnten Versuchsreihen mit Wasser [...] Bei den Werten von $\frac{ud}{\nu}$ zwischen 2000 und 3000 findet ein Übergang von dem einen Gesetz zu dem andern statt." (Blasius, 1912b, p. 37).

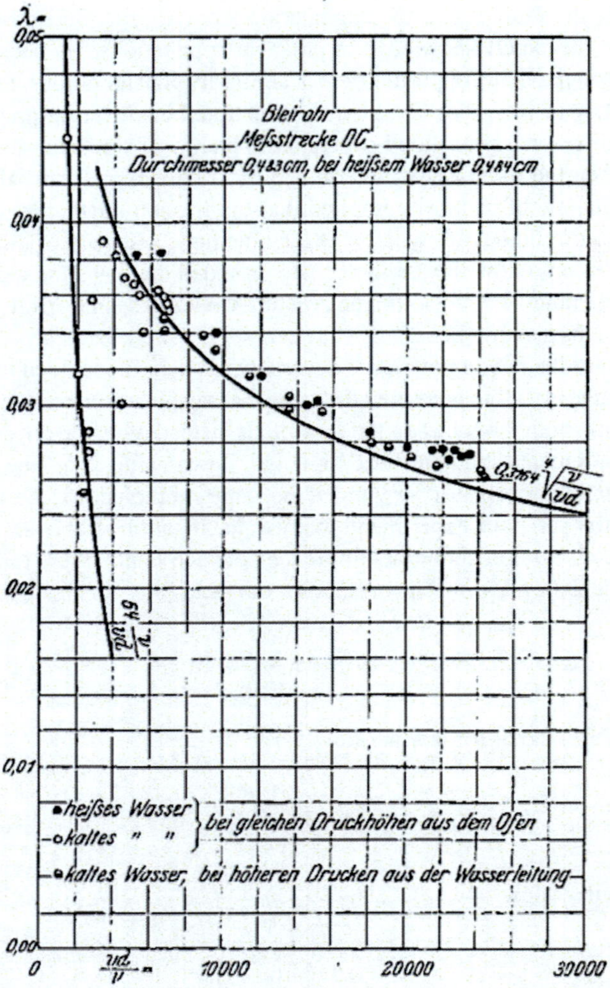

Fig. 1.4 Blasius's diagram for the friction coefficient of laminar and turbulent pipe flow as a function of the Reynolds number (Blasius, 1913, Abb. 13)

the functional dependence of the friction coefficient λ on the Reynolds number did not result from similarity considerations alone; the supposed formulae had to be matched to the experimental data. One third of his final report was dedicated to tables and diagrams with the results of pipe flow measurements. "The most careful and extensive tests on pressure loss in smooth pipes were carried out by the American engineers Saph and Schoder," Blasius acknowledged these experiments as his main source and inspiration. He called his expression for the turbulent friction coefficient the "improved Saph-Schoder formula".[32] But he also performed himself measurements on the pipe flow of water at different temperatures in order to demonstrate how the data from different experiments collapsed when displayed as a function of the Reynolds number (Fig. 1.4). Furthermore, he attempted to trace the origin of discrepancies between measurements of Reynolds and Saph-Schoder. "I therefore checked Reynolds' experiments and found that lead pipes have the same resistance as the Saph-Schoder brass pipes. I therefore suspect that Reynolds has a systematic error in the measurements."[33]

Blasius explicitly confined the validity of his turbulent friction factor to completely smooth pipes. For pipes with a rough inner surface he concluded from the available data that the friction factor could no longer be expressed in terms of the Reynolds number alone. He introduced a parameter ε/d, where ε is the size of the bumbs that protrude out of the surface, but he abstained from deriving a formula. He left the roughness issue with the statement "that in any case no pure power law is possible" for the formula of the friction factor.[34]

An attempt in this regard was made subsequently by Richard von Mises (1883–1953) (Siegmund-Schultze, 2018). A few years later he made himself a name as the founder of the *Zeitschrift für Angewandte Mathematik und Mechanik (ZAMM)* (Siegmund-Schultze, 2020), but at first he regarded himself as a mechanical engineer (Maschinenbau-Ingenieur)—or so he designated himself in September 1913 in the *Elemente der Technischen Hydromechanik* that showcased his familiarity with pipe flow measurements (von Mises, 1914). From Darcy's measurements with rough pipes he derived the following formula for friction factor:

$$\lambda = 0.0024 + \sqrt{\frac{k}{r}} + 0.3\left(\frac{ur}{\nu}\right)^{-1/2}$$

where k is an empirically determined "absolute roughness" and r the radius of the pipe (Mises, 1914, p. 75).

Another report that displayed a wealth of pipe flow data as a function of the Reynolds number was published somewhat later by Thomas E. Stanton (1865–1931) and John Robert Pannell (1885–1921) from the Engineering Department of the National Physical Laboratory (NPL) in Teddington (Stanton and Pannell, 1914). Apparently unaware of Blasius's papers they offered similar diagrams that showed how the data of quite different experiments about the flow of water and air collapsed when displayed as a function of the Reynolds number (Fig. 1.5). Among hydraulic engineers such plots became labelled as Stanton-diagrams (Steen and Brutsaert, 2017, p. 580).

---

[32] "verbesserte Formel von Saph-Schoder" (Blasius, 1913, p. 29), see also (Steen and Brutsaert, 2017, p. 576).

[33] "Ich habe infolgedessen die Reynoldsschen Versuche nachgeprüft und gefunden, daß auch Bleirohre denselben Widerstand wie die Saph-Schoderschen Messingrohre haben. Ich vermute demnach bei Reynolds einen systematischen Fehler der Messungen." (Blasius, 1913, p. 14).

[34] "... dass jedenfalls kein reines Potenzgesetz in Frage kommt..." (Blasius, 1913, p. 21).

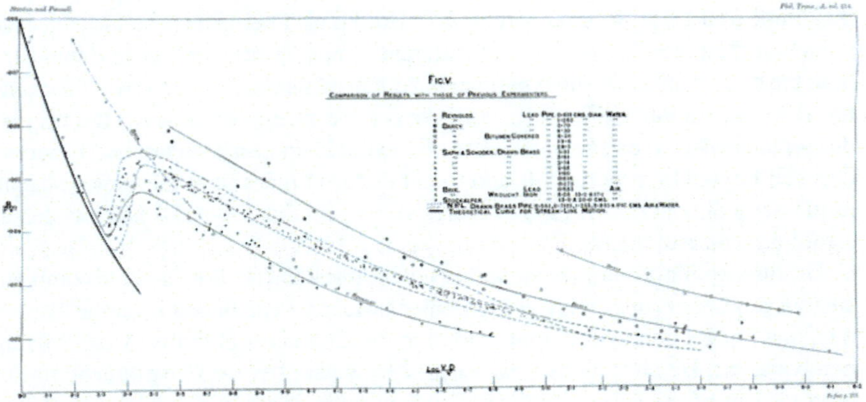

Fig. 1.5 Stanton and Pannell compared the pipe flow data from many different experiments by displaying the friction coefficient as a function of the Reynolds number. The curve for "streamline motion" at low Reynolds numbers, i.e. laminar flow, was clearly distinguished from turbulent flow at higher Reynolds numbers (Stanton and Pannell, 1914, Fig. V)

By the time of World War I, therefore, the evaluation of pipe flow data clearly exposed turbulent flow as a general phenomenon at high Reynolds numbers. Beyond a critical Reynolds number the friction coefficient of smooth pipe flow seemed to follow a law ("Blasius law") for which hydraulic engineering provided abundant empirical evidence—but no theoretical explanation in terms of the fundamental equations of hydrodynamics. This became a major challenge after the war (see Sect. 5.4). The other problem that awaited a physical explanation was the departure from the "streamlime motion", i.e. the transition from laminar to turbulent flow at a critical Reynolds number (see Chap. 8). Both the onset of turbulence and fully developed turbulence at high Reynolds numbers became outstanding challenges for several generations of turbulence researchers in the 20th century.

Chapter 2
Channel Flow

Abstract The relationship between flow velocity, slope and cross-sectional geometry of a channel was crucial for designing water supplies of villages and agricultural areas. As early as in the 18th century hydraulic engineers arrived at a formula with a coefficient that accounted for turbulent channel flow ("Chézy coefficient"). Throughout the 19th century channel flow gave rise to studies on the fringe to turbulence research. Unlike turbulent pipe flow, however, open channel flow turbulence remained difficult to understand from the vantage point of similarity theory.

2.1 Water Management in Early Modern Italy

Like pipe flow the motion of water in rivers and open channels became long ago the subject of practical—and sometimes theoretical—concerns. Half a millennium ago Leonardo da Vinci (1452–1519) penned in his notebook:

> A straight river with equal width, depth, and slope acquires a degree of velocity for each degree of motion. This is evident from the proportion of motion according to which an object, the more it moves in its own natural course, the more it gathers speed, as in any other matter. The straight river as above, although it acquires a grade of velocity for each grade of motion, nevertheless, the more obstacles it encounters on its course, the more it slows down in speed [...] Water has higher speed on the surface than at the bottom. This happens because water on the surface borders on air which is of little resistance, because lighter than water, and water at the bottom is touching the earth which is of higher resistance, because heavier than water and not moving. From this follows that the part which is more distant from the bottom has less resistance than that below. [Quoted in Rouse and Ince (1957, p. 49)].

Leonardo is famous as Renaissance artist, scientist and engineer. He even used the term "turbulence" ("turbolenza") (Macagno, 1986, p. 8) to designate eddying flow.[1] His drawings and notes on the flow of water have been showcased as examples of early fluid mechanics in general, and turbulence in particular. Beyond mere curiosity there were practical challenges that sparked Leonardo's interest: He earned himself the reputation as "Master of Water" from the city of Florence in whose behalf he conceived plans to canalize the river Arno away from Pisa, Florence's enemy. If

[1] For a detailed analysis see Enzo O. Macagno's evaluation of Leonardo's manuscripts on water flow in https://ir.uiowa.edu/.

M. Eckert, *Turbulence—an Odyssey*, History of Physics,
https://doi.org/10.1007/978-3-030-91459-2_2

successful, Pisa would have been deprived from a direct access to the Mediterranean and Florence would have been connected to the sea by a navigable canal. In the course of this and other projects Leonardo must have encountered many problems that were perceived later from the vantage point of open channel hydraulics (Masters, 1998; Marusic and Broomhall, 2021).

River management and the design of canals remained a pressing concern of the Italian city states—but Leonardo soon fell into oblivion as the "Master of Water". His achievements were lost to his successors, not least because he kept them to his private notebooks. Whatever he may have presaged awaited to be rediscovered by and attributed to others. It was left, for example, to Galileo's (1564–1642) pupil Benedetto Castelli (1577–1643) more than a hundred years later to establish the law of continuity for channel flow. He, too, was concerned with river management projects. Unlike Leonardo, however, Castelli did not keep the results from this occupation to himself. He published a treatise *Della misura delle acque correnti* that earned him the reputation as "founder of the Italian hydraulics school" (Rouse and Ince, 1957, p. 59). Another representative of this school from the late 17th century is Domenico Guglielmini (1655–1710). He communicated his observations in treatises on *Aquarum fluentium mensura nova methodo inquisita* and *Della natura dei fiumi*, published in 1690 and 1697. Guglielmini, for example, explicitely addressed the role of the resistance for the establishment of a constant flow velocity when water is discharged from an elevated reservoir in a descending channel. In its descent the water is accelerated downstream, but its velocity "soon reduces to uniformity due to the great resistances that the water encounters in its motion" [quoted in Rouse and Ince (1957, p. 70)]. Like Castelli and other Italian hydraulic engineers from this era, Guglielmini arrived at his conclusions from observations made in the course of one or another water management project. Fuelled by practical needs river and open channel hydraulics ranked high in Italy throughout the 17th century (Maffioli, 1994).

2.2 Chézy's Formula

By the 18th century investigations of open channel flow found an increasing attention also in other countries. In November 1732 Henri Pitot (1695–1771) communicated to the Royal Academy of Science in Paris the description of a "machine for measuring the speed of flowing water and the wake of vessels" that enabled more precise channel flow experiments. Previously the only means for measuring the flow speed had been to observe the passage of a piece of wood or other floating objects within a certain span of time. "Two experiments conducted at the same site of a river often display quite different velocities, because the piece of wood or the ball of wax not always take the same way in the water," Pitot remarked about the erratic results of flow measurements by the old method due to the eddying motions at the surface of a river. Thus the flow in the wake of the pillar of a bridge or at the entrance or exit of a canal remained inaccessible to measurements. Furthermore, it was left to speculation whether the speed towards the bottom increased or decreased. Guglielmini and others had assumed that the water on the bottom was forced to flow faster because the weight of the upper layers pressed the lower ones. Pitot opposed this view because it ignored the friction:

> Since the water flow is slowed down so considerably by the friction of the bed and the borders, it is natural to think that the flow near the bottom is slower than that at the surface. All these questions, equally useful and curious, can be clarified on the field with great ease by means of the instrument that I propose, since the operation is as simple as plunging a stick into the water and removing it. With this machine one can measure the right amount of water speed at any depth, as easily as at the surface.[2]

The instrument entered the history of fluid mechanics as "Pitot tube". It was based on false conceptions of fluid percussion ["Pitot fell far short of [Daniel] Bernoulli's correct introduction of the piezometric principle" (Rouse and Ince, 1957, p. 116)] and left room for improvement before it could be employed for precise measurements (Brown, 2003), but this did not prevent its inventor from a successful career. Pitot embodied the wide range of interests and affiliations of an academician in 18th century France. He also designed bridges and published papers on geodesy, astronomy and mathematics (de Fouchy, 1774). Among other employments for practical works he served as superintendent of the Canal du Midi—by his time *the* world renowned showcase of hydraulic engineering. Contemporaries like Pierre du Buat praised him for his observation that the flow velocities in an open channel or river "always diminish from the surface to the bottom", as du Buat remarked in the final chapter of his *Principes d'Hydraulique* that focussed on "Considerations on some objects regarding the resistance of fluids".[3]

Although Pitot's instrument did not yet enable precision measurements of velocity profiles, it was sufficient to conclude that the resistance exerted by the bottom and the borders of a canal retarded the accelerating motion by gravity due to the slope of the canal into a flow at constant mean velocity. In 1775 Antoine de Chézy (1718–1798), "Ingénieur des Ponts et Chaussées" and "Ingénieur général du pavé de Paris", turned this conclusion into the quest "to find a formula for the uniform velocity that the water will have in a ditch, or in a regular canal of which the slope is known."[4]

[2] "Puisque les eaux sont rallenties si considérablement par les frottements de leurs lits et des bords, il est naturel de penser que celles qui sont près du fond sont plus rallenties que celles qui sont à la surface. Toutes ces quéstions également utiles et curieuses, peuvent être éclaircies sur le camp avec une grande facilité au moyen de l'instrument que je propose, puisque l'opération en est aussi simple que celle de plonger un bâton dans l'eau et de le retirer. Par cette machine on mesurera la juste quantité de la vitesse des eaux à telle profondeur qu'on voudra, et cela aussi aisement qu'á leur surface." (Pitot, 1735, p. 366).

[3] "elles vont toujours en diminuant de la surface au fond"; the chapter is titled "Considérations sur quelques objets relatifs à la résistance des fluides: conclusions de cet Ouvrage". In the same vein du Buat remarked that Pitot's instrument did not yield exact results and suggested an improvement "de le rendre d'un usage plus général, pour connoître les vitesses réelles d'une rivière en un lieu quelquonque de sa section." (Buat, 1786, II, pp. 331, 333).

[4] "Formule pour trouver la vitesse constante que doit avoir [l'eau dans] une rigole ou un canal dont la pente est donnée" (Mouret, 1921, p. 247). The report was lost, but Du Buat mentioned Chezy's investigations in his *Principes d'Hydraulique* (Buat, 1786, I, 316–322). The manuscript upon which the report was based ("dossier 847 Ms. 1915") was preserved in the collection of manuscripts at

Chézy's investigation was part of a project to improve the Paris water supply by a canal that would convey water from the river Yvette to the city. In his report, Chézy described the challenge in the following way:

> After having designed a ditch or channel, and having adjusted and regulated its slope, it is very interesting to know if the section of this channel will be sufficient to conduct the water which is to flow in it. To know this, it is necessary to know the speed which the water will have in the ditch, which we will suppose to have a uniform slope.[5]

Chézy assumed that the resistance is proportional to the square of the mean velocity u and acts along the perimeter p of the cross section of the flow a that is in contact with the bottom and the borders of the channel. He neglected the air resistance at the surface. Then the resistance would be proportional to $u^2 p$ and the driving force due to gravity ah, where h was a measure for the slope. Uniform flow implied that the proportion $u^2 p/ah$ is constant. Thus, the quantities of two different channels are related via

$$\frac{U^2 P}{AH} = \frac{u^2 p}{ah},$$

where the upper case corresponds to a channel where all quantities, including U, are known, and the lower case to a channel where only p, a, h are given. Then the unknown velocity u resulted from $U^2 Pah = u^2 pAH$, or

$$u = U\sqrt{\frac{ahP}{AHp}}.$$

Chezy's formula relied on the comparison with a canal whose walls had the same roughness and whose other quantities were known from measurement. For the Yvette project he used in one experiment a short piece of the projected new canal as a reference, in another he used measurements from the river Seine for comparison. "For the present, and by means of the preceding formula, we are enabled to compute the velocity of the water in a projected channel, from either one of the two observations above given."[6]

Chézy's formula may be written in modern form as

---

the library of the Ecole des Ponts et Chaussées. It was made known in English (Herschel, 1897) before the French original was published, together with a history of the Chézy formula throughout the 19th century (Mouret, 1921). The broader community of hydraulicians learned about it only from (Rouse and Ince, 1957, pp. 117–120).

[5] "Après avoir projeté une rigole, et en avoir bien ménagé et réglé la pente, il est très intéressant de savoir si cette rigole sera suffisante pour l'eau qui doit y couler. Pour cela, il faut connaître la vitesse que l'eau pourra avoir dans le cours de cette rigole, qu'on suppose d'une pente uniforme." (Mouret, 1921, p. 247). English translation in (Herschel, 1897, p. 363).

[6] "On peut, à présent, par le moyen de la formule précédente, et de l'une ou de l'autre des deux observations, trouver la vitesse de l'eau d'une rigole projetée" (Mouret, 1921, p. 245). English translation in (Herschel, 1897, p. 367).

$$u = C\sqrt{Ri},$$

with $C = U\sqrt{\frac{P}{AH}}$, $R = \frac{a}{p}$ = hydraulic radius and $i = h$ = hydraulic gradient. But it should be noted that Chézy did not regard C as a constant; he perceived his formula only as a tool for designing a new canal by comparison with a reference.

Chézy's formula served as a bone of contention whenever hydraulic engineers attempted to account for friction in channel flow. Pierre Simon Girard, another engineer from the École des Ponts et Chaussées who was also involved in practical channel design, observed that Chézy's approach was not appropriate if the flow velocity in the channel was very small. He referred to Charles Coulomb whose famous torsional-balance experiments had given rise to investigations on the drag of very slowly moving bodies. In a treatise published in 1800 Coulomb had concluded that there were two mechanisms involved: the mutual interaction of fluid particles and their collision with obstacles at the surface of the solid body. The former resulted in a linear, the latter in a quadratic dependence on the velocity. At very slow motion Coulomb observed that "the part of the resistance which we found to be proportional to the velocity is due to the mutual adherence of the molecules, not to the adherence of these molecules with the surface of the body."[7] In the same vein Girard argued that the retarding force at small velocity was due to the mutual adherence of the fluid molecules where the wetted surface at the wall was covered by a fluid layer:

> Due to the effect of the adhesion of the fluid to the walls of the channel in which it moves, it happens that a more or less thick layer of this fluid remains attached to its walls, so that the rest of the fluid moves by sliding on this layer, the thickness of which is constant, if the walls are homogeneous, and which, consequently, presents to the moving mass the same asperities as the walls themselves.
>
> Any section of the current is therefore held in the position it currently occupies, by the force with which all the molecules of its perimeter adhere to the fluid layer that wets the walls.[8]

Girard assumed that this friction was proportional to the velocity. Another retarding force was due to the collision of the fluid particles with the protrusions from the channel bed ("la force retardatrice provenant du choc contre les aspérités du lit") for which Girard assumed the same as Chézy and Coulomb, a dependence on the square of the velocity. Thus he arrived at a formula with a linear and quadratic term (Girard, 1803, p. 38).

Gaspard de Prony, director of the à cole des Ponts et Chaussées and a former student of Chézy, further extended Chézy's and Girard's concept. He praised Girard's

---

7 "... que la partie de la résistance que nous avons trouvée proportionelle à la simple vîtesse, est due à l'adhérence des molécules du fluide entre elles, et non à l'adhérence de ces molécules avec la surface du corps." (Coulomb, 1800, p. 287), translated in (Darrigol, 2005, p. 105).

8 "Par l'effet de l'adhérence du fluide aux parois du canal dans lequel il se meut, il arrive qh'une couche plus ou moins épaisse de ce fluide reste attachée à ses parois, de sorte que le reste du fluide se meut en glissant sur cette couche, dont l'épaisseur est constante, si les parois sont homogènes, et qui, par conséquent, présente à la masse mobile les mêmes aspérités que les parois elles-mêmes.

Une section quelquonque du courant est donc retenue dans la position qu'elle occupe actuellement, par la force avec laquelle toutes les molécules de son périmètre adherent à la couche fluide qui mouille les parois." (Girard, 1803, p. 36).

introduction of a boundary layer ("la couche d'eau adhérente à cette paroi") in addition to the roughness as cause of friction, but unlike Girard he did not assume that both depended in the same way on the wetted perimeter. The quadratic term of the resistance caused by the wall roughness "must be supposed to be greater than that which retains the fluid molecules to each other," he argued.[9] Thus he arrived at a formula with two different coefficients for the linear and the quadratic term, respectively. The same type of formula accounted for pipe flow (see Sect. 1.3).

2.3 Precise Channel Flow Measurements

Henceforth, channel and pipe flow were often made subject of common theoretical elaboration. Prony's formula, however, suggested more common ground than the experiments on both flows. The motion of water in open channels involved a variety of surface phenomena (waves, backwater, hydraulic jumps) that were absent in pipe flow. Progress of channel flow theory was even more dependent on precise measurements than that of pipe flow. The perfection of the Pitot tube (Brown, 2003) enabled Henry Darcy in the 1850s to improve Prony's formula for pipe flow (see Sect. 1.5)—it paid off even more when it was applied for channel flow measurements. "Perfections brought by Darcy to this instrument are considerable", a colleague from the École des Ponts et Chaussées praised this achievement. "The administration has already made a number of these instruments some of which are placed in the precision instrument deposit at l'École des Ponts et Chaussées, and others sent to state engineers to be used in operations that demand exact measurements of the water velocity." [quoted in Brown (2003, p. 22)]

After his pipe flow investigations Darcy started a series of experiments on channel flow, but poor health prevented him from pursuing these studies so that it was left to his collaborator, Henri Bazin, to accomplish this project. The results were published after Darcy's death under their common authorship in a voluminous treatise of which the first part was dedicated to their investigations on a test channel diverted from the Bourgogne canal at Dijon, and the second part on wave propagation and backwater phenomena (Darcy and Bazin, 1865). In the test channel they modified both the cross-section (rectangular, trapezoidal, semi-circular) and the boundary (wood, brick, cement, gravel, rock). As a result they left little hope for a general formula that would be able to account for all aspects, in particular the type of boundary:

> The slightest alteration of the wall is sufficient to modify the flow in an appreciable way; the analysis cannot, of course, take into account these fortuitous alterations, which can be

[9] "Il me semble qu'en raisonnant d'après l'hypothèse de l'égalité d'adhérence des aspérités à la paroi et des molécules fluides entre elles, il faudrait de plus énoncer que les aspérités disséminées sur la paroi ne sont pas contiguës les unes aux autres [...] Je pense encore, si l'on admet, pour expliquer la partie de la résistance proportionnelle au carré de la vîtesse, les obstacles dus à des aspérités disséminées sur la paroi, que l'adhérence de ces aspérités à cette paroi doit être supposée plus grande que celle qui retient les molécules fluides les unes aux autres." (Prony, 1804, pp. 53–54).

> produced by a thousand different causes, and, therefore, the search for a rigorous formula becomes, from a practical point of view, completely illusory.[10]

In particular they dismissed Prony's two-coefficient-formula

$$Ri = aU + bU^2,$$

where R is the hydraulic radius (cross section divided by wetted perimeter), i the hydraulic gradient, U the mean velocity averaged over the cross section, and a, b coefficients which Prony had assumed to be constant. Bazin suggested instead the Chézy-type formula

$$Ri = (\alpha + \frac{\beta}{R})U^2,$$

with the remark that "the coefficients α and β are certainly not constants but valid only for one and the same type of boundary".[11] Thus they recovered Chézy's coefficient (see above) as a quantity that accounted for the roughness of the channel boundary.

Bazin also determined the velocity distribution over the cross section of the test channel. These measurements were not only conducted for the precise determination of the discharge and the mean velocity of the flow in the channel, but also with regard to theoretical speculations on the friction of fluid molecules against each other. Bazin cited the view of Saint-Venant whereas the flow through narrow cross sections occurs in parallel threads with a gradual change of velocity, but that the flow through wider cross sections is different. Saint-Venant opined that in this case

> the threads do not run parallel to each other with speeds regularly graduated from one to the other, and that breaks, swirls and other complicated or oblique movements, which must have a great influence on the size of the friction, form and develop more in large sections.[12]

Bazin found large differences in the velocity distribution for channels with different cross sections, in particular near the surface where he observed frequent "tumultuous movements" ("mouvements tumultueux") (Darcy and Bazin, 1865, p. 25). He considered such erratic fluid motions inaccessible to hydrodynamic theory, at least

---

[10] "La plus légère altération de la paroi suffit pour modifier l'écoulement d'une manière appréciable; l'analyse ne saurait évidemment tenir compte de ces altérations fortuites, qui peuvent être produites par mille causes diverses, et, dès lors, la recherche d'une formule rigoureuse devient, au point de vue pratique, complètement illusoire." (Darcy and Bazin, 1865, p. 125).

[11] "Les coefficients α et β ne sont pas tout à fait constants, il est vrai pour une même nature de paroi." (Darcy and Bazin, 1865, p. 125).

[12] "... ce qui s'explique jusqu' à un certain point en remarquant que les filets ne marchent pas parallèlement entre eux avec des vitesses régulièrement graduées de l'un à l'autre, et que les ruptures, les tourbillonnements et les autres mouvements compliqués ou obliques, qui doivent beaucoup influer sur la grandeur des frottements, se forment et se développent davantage dans les grandes sections." [Saint-Venant, quoted in Darcy and Bazin (1865, p. 29)].

for the time being. “Maybe such a delicate part of science must long remain in the realm of experiment.”[13]

Bazin’s measurements turned open channel flow into a flourishing field of hydraulic engineering, not only for the Corps des Ponts et Chaussées in France. In 1869 two Swiss engineers, Emile Oscar Ganguillet (1818–1894) and Wilhelm Rudolph Kutter (1818–1888), published a generalized version of Chézy’s formula on the basis of Bazin’s results—and this was just the beginning of a new quest to bring the variety of ever more precise channel flow measurements under the umbrella of a common formula for engineering purposes (Rouse and Ince, 1957, pp. 177–180; Hager, 1994). Thus open channel flow investigations broadened the empirical knowledge of turbulence, although turbulence was rarely addressed explicitly as the culprit of the involved difficulties.

2.4 Boussinesq’s Friction Coefficient ε

One direct approach on turbulence inspired to a large extent by channel flow investigations was launched by Joseph Boussinesq (see Sect. 1.6). In his *Essai sur la Théorie des Eaux Courantes* Boussinesq included turbulent fluctuations into the Navier-Stokes equations by assuming that these fluctuations cause an additional friction to the molecular viscosity of the fluid. Thus he introduced a variable friction coefficient ε which varied from place to place in a turbulent flow dependent on the local velocity fluctuations (Boussinesq, 1877, pp. 24–46). The origin of this concept may be traced back to speculations about an “effective viscosity” suggested by Boussinesq’s teacher Saint-Venant (Darrigol, 2005, Sect. 6.2.2; Darrigol, 2017, p. 11). Boussinesq ventured to derive approximate formulae for simple cases such as the flow in a wide rectangular channel (“quand la section est rectangulaire très-large”). He obtained the variation of the mean flow velocity from the bottom to the surface in good agreement with Bazin’s measurements; furthermore he derived an expression that related the velocity averaged over the cross-section to the slope of the channel (Boussinesq, 1877, pp. 72–88; Darrigol, 2017, pp. 11–12).

Despite some criticism [e.g. (Lorentz, 1897)] Boussinesq’s approach was perceived by theorists and practitioners as a promising way for rendering turbulence such as observed in open channel flow accessible to theory. In 1904 the Göttingen mathematician Felix Klein (1849–1925) presented Boussinesq’s effort in a seminar on hydrodynamics as a role model how theorists should approach hydraulic engineering. One talk of Klein’s seminar was dedicated to turbulence—with the explicit aim to demonstrate the agreement of Boussinesq’s theory with Bazin’s experiments. The lesson was regarded important enough for publication in the *Zeitschrift für Mathematik und Physik*:

[13] “Peut-être cette partie si délicate de la science doit-elle rester longtemps encore dans le domaine de l’empirisme.” (Darcy and Bazin, 1865, p. 30), translated in (Darrigol, 2005, p. 233).

> In the following we assume with Boussinesq for turbulent motion the validity of the hydrodynamic equations involving a variable friction coefficient ε. By ε we mean a quantity proportional to the energy of the turbulent motion and we will therefore take the liberty of referring to ε briefly as "turbulence". We see our further task in deriving the values of ε from the existing observations on the basis of the differential equation [of Boussinesq], and in this way to obtain information about the behaviour of the turbulence. [...]
>
> After all, the results suggest that an idea of the turbulence conduction has been gained here, which can serve as a first clue for further investigations.[14]

The same lesson was spread in 1905 in an encyclopedia article on hydraulics. The author made "Boussinesq's approach regarding friction" the subject of an entire section based on "the idea that the flow velocity is the average of continuously changing pulse velocities". These "pulsations" produced

> a permanent mixing, i. e. an exchange of velocities (similar to the gaseous friction according to the kinetic gas theory). Therefore ε is a measure of the local, from place to place varying "turbulence". Conclusions about the magnitude of the turbulence can now be drawn from Bazin's experiments, for which on the other hand there is a certain probability from the outset.[15]

The author of this encyclopedia article was Philipp Forchheimer (1852–1933), an Austrian engineer and professor of hydraulics and bridge building at the Technical University in Istanbul. It is characteristic for the gulf between hydraulics and hydrodynamics around 1900 that the latter was reviewed in the same encyclopedia by Augustus Edward Hough Love (1863–1940), a mathematician professor of Natural Philosophy at Oxford University (Love, 1901a, b). Not surprisingly Love bestowed little attention to channel flow.

Boussinesq's approach was not confined to open channel flow; it yielded comparable results also for pipe flow (Forchheimer, 1905, pp. 349–350). Yet the very idea behind ε addressed fully developed turbulence. The onset of turbulence escaped Boussinesq's theory. It was open to inquest whether the transition from laminar to turbulent flow as analysed by Reynolds suggested similar conclusions for open channel flow.

---

[14] "Im folgenden nehmen wir jedenfalls mit Boussinesq für die turbulente Bewegung die Gültigkeit der hydrodynamischen Gleichungen bei variablem Reibungskoeffizienten ε als gegeben an. Unter ε haben wir dabei eine der Energie der turbulenten Bewegung proportionale Größe zu verstehen und wollen uns daher erlauben, ε kurz als 'Turbulenz' zu bezeichnen. Wir sehen unsere weitere Aufgabe darin, aus den vorhandenen Beobachtungen auf Grund der Differentialgleichung (2) die Werte von ε abzuleiten und auf diese Weise Aufklärung über das Verhalten der Turbulenz zu gewinnen. [...] Immerhin lassen die vorhin gewonnenen Resultate darauf schließen, daß hier eine Vorstellung über die Turbulenzleitung gewonnen ist, die als erster Anhalt bei weiteren Untersuchungen dienen kann." (Hahn et al., 1904, pp. 414–415, 426).

[15] "Diese verstärkte Reibung wird durch ein fortwährendes Durcheinandermischen der Flüssigkeit, also ein Austauschen der Geschwindigkeiten bewirkt (ähnlich wie die Gasreibung nach der kinetischen Gastheorie). Daher giebt dann ε das Maass der an der betreffenden Stelle herrschenden, von Stelle zu Stelle sich ändernden 'Turbulenz'. Über die Grösse der Turbulenz lassen sich nun gerade aus den Bazin'schen Versuchen Schlüsse ziehen, für die andererseits auch von vorneherein eine gewisse Wahrscheinlichkeit spricht." (Forchheimer, 1905, p. 347).

2.5 Onset of Channel Flow Turbulence

In 1909 Ludwig Hopf (1884–1939) attempted to fill this gap with an experimental investigation on "turbulence in a river," a subject that was suggested to him by Sommerfeld as part of his doctoral dissertation. Such a study appeared "tempting" not only because it was unprecedented but also because the "behaviour of the free surface" would provide new insights into the transition process. "There is under certain circumstances (small depth, slow flow velocity) a stable laminar motion of the kind of Poiseuille's law, which becomes unstable beyond a critical limit and gives rise to another form of motion," Sommerfeld summarized the goal of Hopf's experiments.[16] Hopf used a straight 5.2 cm wide brazen water channel and varied the flow speed by increasing the channel's inclination. He also modified the viscosity by adding sugar. "I still remember how I brought him a whole sack-full of sugar so he could make a sugar solution," Sommerfeld's assistant recalled many years later Hopf's experiment on turbulence.[17] As a result, Hopf observed a sharp transition at a Reynolds number

$$R = \frac{uh\rho}{\mu} = \frac{Q}{2b}\frac{\rho}{\mu} = 330,$$

where u is the mean flow velocity, h the depth of the fluid in a channel of width b, Q the discharge, ρ and μ the density and viscosity of the fluid (Hopf, 1910, pp. 7–8, 18–19). "Within the limits of errors Reynolds' law may therefore be confirmed also for rivers", Hopf concluded. "Variation of slope is equivalent with variation of velocity and depth."[18] However, the transition to turbulence in Hopf's "river" was not only dependent on the Reynolds number but also on the formation of capillary waves (Hopf, 1910, p. 27). Unlike Reynolds' demonstration with pipe flow the corresponding effort with open channel flow proved elusive for a closer investigation of the onset of turbulence.

There was also a marked difference between pipe flow and open channel flow with regard to fully developed turbulence: whereas the friction factor for turbulent pipe flow could be expressed in terms of the Reynolds number ("Blasius' law"), the Reynolds number dependence of fully turbulent open channel flow remained a subject of debate well into the twentieth century. The Reynolds similitude is characterized by the ratio of inertial and viscous forces, but for the flow in large channels another

[16] "Auch hier ergiebt sich unter gewißen Umständen (kleine Tiefe, kleine Geschwindigkeit) eine stabile geradlinige Strömung im Sinne der Poisseulle'schen Gesetze, von einer gewißen kritischen Grenze ab ein Instabilwerden dieser Strömung und Umschlagen in eine andere Bewegungsform." Sommerfeld's report to the philosophical faculty of Munich University, 5 July 1909. Munich University Archive (OC I 35 p).

[17] Peter Debye (1884–1966), interviewed by T. S. Kuhn und G. Uhlenbeck, 3 May 1962. AIP. Available online at http://www.aip.org/history/ohilist/4568_1.html.

[18] "Das Reynoldssche Gesetz kann also an Flüssen von verschiedenem Gefälle innerhalb der Fehlergrenze bestätigt werden. Variation des Gefälles ist gleichbedeutend mit Variation der Geschwindigkeit und der Tiefe." (Hopf, 1910, p. 21).

similitude consideration plays a role: the ratio of inertial forces to the weight of the fluid. This “Froude similitude” was discovered in model experiments on the drag of ships (see Sect. 3.4). Its pertinence for turbulence in open channels remained even more controversial than that of the Reynolds similitude. From the perspective of similarity theory and model experiments, therefore, open channel flow did not lend itself as easily as pipe flow to the derivation of more general laws on turbulence.

Chapter 3
Drag

Abstract Controlling the flow of water in rivers and channels or through long conduits of pipes was not the only practical concern that gave rise to studies on turbulence centuries ago. Another challenge that called for a mastery of eddying flow arose in the context of naval architecture, water power technology, ballistics and other areas where the drag of bodies in the stream of water or air played a role. Crafting water wheels was perhaps the earliest of these challenges. The flow around the blade of a water wheel was subject of controversies from the 17th to the 20th century. In the 18th and 19th centuries drag studies were performed on vessels in towing experiments. By the 20th century the drag measurements of streamlined and blunt bodies in wind tunnels marked the beginning of a new branch of turbulence research.

3.1 Hog's Bladders in St. Paul's Cathedral

Throughout the centuries the behaviour of a body in a fluid was a fundamental theoretical problem and an obvious practical concern. The motion of celestial bodies, ships, projectiles, and other phenomena involved conceptions of fluid dynamics. Galileo Galilei, with whom we associate the revolutionary turn from the medieval philosophy to the "New Science" of motion, conveyed the view that the motion of a body in a fluid meets "with a resistance which is proportional to the rapidity with which the medium must give way to the passage of the body" (Galilei, 1950, p. 74). Isaac Newton (1642–1726) regarded the relation "that the resistance is in the ratio of the velocity" as "more a mathematical hypothesis than a physical one". Instead of a linear relation Newton suggested a square law for the drag of a body in a flow. "In mediums void of all tenacity, the resistances made to bodies are in the duplicate ratio of the velocities" (Newton, 1729, Book II, Section I, Proposition IV, Scholium). In another proposition he arrived at the theorem "that the greater parts of the systems are resisted in a ratio compounded of the duplicate ratio of their velocities, and the duplicate ratio of their diameters, and the simple ratio of the density of the parts of the systems" (Newton, 1729, Book II, Section VII, Proposition XXXIII, Theorem XXVII).

M. Eckert, *Turbulence—an Odyssey*, History of Physics,
https://doi.org/10.1007/978-3-030-91459-2_3

Newton attempted to confirm his resistance theorem with experiments. "In order to investigate the resistances of fluids from experiments, I procured a square wooden vessel [...] this I filled with rain-water: and having provided globes made up of wax, and lead included therein, I noted the times of descents". Thus Newton described the beginning of a series of experiments on fluid resistance. He concluded from these measurements "that the theory agrees with the phenomena of bodies falling in water; it remains that we examine the phenomena of bodies falling in air" (Newton, 1729, Book II, Section VII, Proposition XL, Experiment 12).

Some of these experiments were performed in St. Paul's cathedral. "From the top of St. Paul's church in London in June 1710 there were let fall together two glass globes, one full of quicksilver, the other of air," he started to describe an astonishing experiment. The two spheres traversed a height of 220 English foot (67 m) before they shattered into pieces on the cathedral's floor. They were released by a sophisticated trapdoor-mechanism which ascertained their simultaneous begin of fall. The experiment was repeated several times with varying weights. Another series of free fall experiments in St. Paul's cathedral was conducted with hog's bladders:

> Anno 1719 in the month of July, Dr. Desaguilers [1683-1744] made some experiments of this kind again, by forming hogs bladders into spherical orbs; which was done by means of a concave wooden sphere, which the bladders, being wetted well first, were put into. After that, being blown full of air, they were obliged to fill up the spherical cavity that contained them; and then, when dry, were taken out. These were let fall from the lantern on the top of the cupola of the same church; namely from a height of 272 ft [...] the bladders did not always fall directly down, but sometimes fluttered a little in the air, and waved to and fro as they were descending [...] Our theory therefore exhibits rightly, within a very little, all the resistance that globes moving either in air or in water meet with; which appears to be proportional to the densities of the fluids in globes of equal velocities and magnitudes. (Newton, 1729, Book II, Section VII, Proposition XL, Experiment 14)

Newton's square law for the resistance of spheres proportional to $\rho D^2 u^2$, where D is the diameter of a sphere moving at constant velocity u in a fluid of density ρ, looks like a valid description of the drag of bodies in a flow. Although Newton's experiments seemed to corroborate his theory, Newton's corpuscular concept of fluids (Newton, 1729, Book II, Section VII, Propositions XXIII, XXXIII and XXXV) associated his resistance law with false conclusions. If the medium is perceived as a system of independent particles, the drag exerted by the collisions of these particles would depend only on the front of a moving body and ignore the modification of the flow by the body's shape in the rear. This flaw was not resolved until the 20th century (Flachsbart, 1932, pp. 22–23).

> Newton's alternating regard of a fluid as a 'rare medium, consisting of [nearly static] equal particles freely disposed at equal distances from each other,' or as 'compressed' [i. e., continuous], while sometimes praised as evidence of great physical insight, did not lead to trustworthy conclusions. (Truesdell, 1954, p. XIII)

This was the way how Newton's concept of fluids was evaluated in the history of rational fluid mechanics. "Book II of Newton's *Principia*, nearly a third of the whole, concerns fluids. Almost all of the results are original, and but few correct"

(Truesdell, 1954, p. XII). Among the few correct statements was the principle of reciprocity which assumed that a body moving at a constant speed in a fluid at rest is opposed by the same resistance as the body at rest in a flow of the same speed. In Newton's words it was the same "whether the body moves in a quiescent medium, or whether the particles of the medium impinge with the same velocity upon the quiescent body" (Newton, 1729, Book II, Section VII, Propositions XXXIV, Theorem XXVIII). Despite Newton's false argument of impinging particles, this theorem provided the basis for investigations of the drag of moving bodies in fluids at rest or of bodies at rest in a flow, such as in water channels and wind tunnels.

3.2 Water Wheels and Wind Mills

Drag investigations, therefore, were not confined to moving bodies in a quiescent fluid. Moving water was supplied to monastries, villages and cities from antiquity to the middle ages in pipes and canals for a host of uses (Magnusson, 2001, Chap. 3). Water-driven machinery in mills was set in motion by the impulse of running water on the blades of a water wheel suspended in a stream ("undershot") or by the drop of water guided on the blades from above ("overshot"). Once more Leonardo da Vinci left notes that witness his practical and scientific interest in such machinery. He even sketched a device for measuring the impulse of a stream of water against a blade, but he did not report results from actual experiments (Reynolds, 1983, p. 199).

The earliest measurements of the force on a blade in a stream of water were reported in Edme Mariotte's *Traité du Mouvement des Eaux et des Autres Corps Fluides*. The experiment compared the force exerted on a quadratic plate at different flow velocities with the respective weights that balanced the "choq" of the flow. Mariotte found that this force was proportional to the square of the flow velocity. "Here is how we can calculate the force of the wheels of the mills at the Seine", he applied this result to a practical case. At the flow velocity of the Seine (4 ft per second) the blades of these mills (5 by 2 ft) could balance a weight of 210 pounds.[1]

Mariotte's experiments suggested the "rule" that flows of the same size but different velocity "balance by their impulse weights which are related to another as the square of their velocities." Another rule stated that the impulse was proportional to the area exposed to the flow (see Fig. 3.1). Mariotte applied his rules not only to water wheels like those from the mills at the Seine, but also to wind mills. "The force of the impulse of the wind against the blades of a wind mill is of the same kind."[2]

---

[1] "Voicy comme on peut calculer la force des roues des moulins de la Seine." (Mariotte, 1686, pp. 221–222).

[2] "Les jets d'eau d'égale largeur qui ont des vitesses inégales, soutiennent par leur choq des poids qui sont l'un à l'autre en raison doublée de ces vitesses. [...] Les jets d'eau de même vitesses et de differentes ouvertures soutiennent des poids par leur choq qui sont l'un a l'autre en raison doublée des ouvertures. [...] La force du choq du vent contre les ailes d'un moulin à vent se trouve en cette sorte." (Mariotte, 1686, pp. 205, 212, 222).

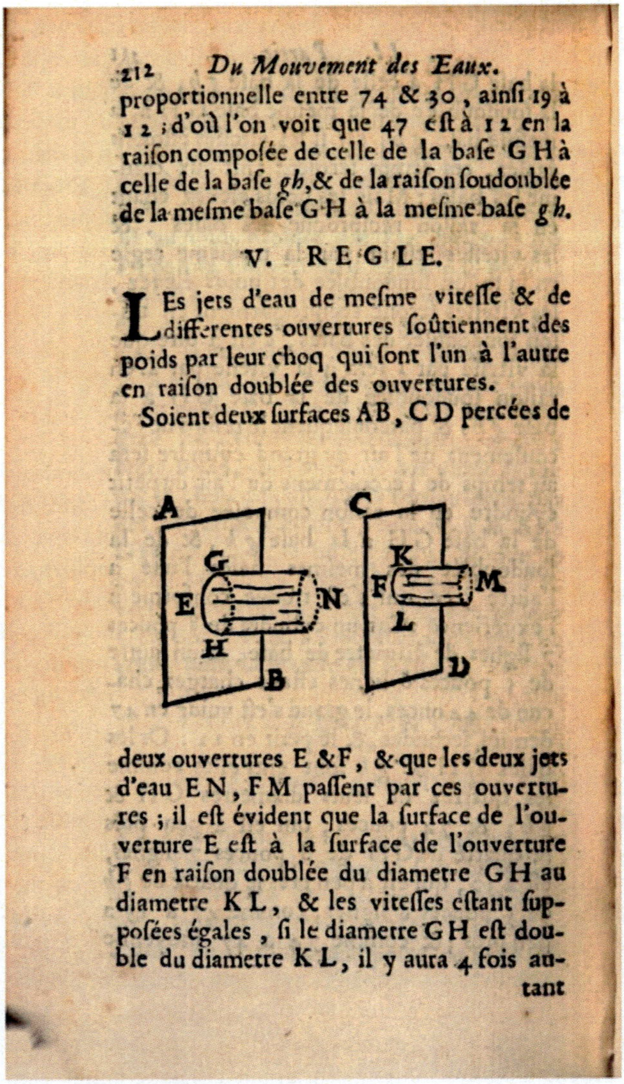

212 *Du Mouvement des Eaux.*
proportionnelle entre 74 & 30, ainſi 19 à 12 ; d'où l'on voit que 47 eſt à 12 en la raiſon compoſée de celle de la baſe GH à celle de la baſe *gh*, & de la raiſon ſoudoublée de la meſme baſe GH à la meſme baſe *gh*.

V. REGLE.

Les jets d'eau de meſme viteſſe & de differentes ouvertures ſoûtiennent des poids par leur choq qui ſont l'un à l'autre en raiſon doublée des ouvertures.

Soient deux ſurfaces AB, CD percées de

deux ouvertures E & F, & que les deux jets d'eau EN, FM paſſent par ces ouvertures ; il eſt évident que la ſurface de l'ouverture E eſt à la ſurface de l'ouverture F en raiſon doublée du diametre GH au diametre KL, & les viteſſes eſtant ſuppoſées égales, ſi le diametre GH eſt double du diametre KL, il y aura 4 fois autant

tant

Fig. 3.1 Edme Mariotte's rule for the impact of jets on plates (Mariotte, 1686, p. 212)

By the early 18th century the force exerted on the blades of water wheels or wind mills became subject of increasing concern. Mariotte's pupil, Philippe de La Hire, extended the investigations of his mentor by a study of the force required to pull a boat with area A of the bow at the velocity u against a stream of velocity U. Antoine Parent (1666–1716), another member of the Paris academy, turned La Hire's investigations into a theory of the power output of water wheels. Unlike the quest for a general theory which did not yield satisfying results, the demand for effective mills gave rise to sophisticated experiments. In 1759, John Smeaton (1724–1792), "one of the most influential figures in the history of water-power technology" (Reynolds, 1983, p. 223), presented to the Royal Society the results of "An experimental enquiry concerning the natural powers of water and wind to turn mills, and other machines, depending on a circular motion" (Smeaton, 1759) that surpassed most preceding experiments by precision and significance. His results proved Parent's theory about the efficiency of undershot water wheels wrong and corroborated earlier speculations about the superior performance of overshot water wheels. Unlike the action of gravity which can be transferred almost completely to the blades of the overshot water wheel, the undershot water wheel is driven by the "impulse or collision" of "nonelastic bodies" of which only a part is communicated to the blades, "the other part being spent in changing their figure in consequence of the stroke" (Smeaton, 1759, p. 130).

Besides the efficiency of water wheels, Smeaton also investigated that of windmills. Qualitatively, Smeaton imagined that the power of the wind is communicated to the sails of the windmill in the same way (by impulse) as that of the water to the blades of an undershot water wheel. The experimental apparatus for measuring this force, however, required a different approach: Smeaton used a whirling arm which could be rotated at a given speed so that the sails mounted at its extremity moved at a given speed against still air (Fig. 3.2). Smeaton found that the load required for balancing the force of the wind "is nearly, but somewhat less than, as the square of the velocity of the wind, the shape and position of the sails being the same." (Smeaton, 1759, p. 154).

The experiments with waterwheels and windmills were conducted by engineers concerned with the use of water- and airflow for driving machinery. Despite widely different applications they amounted to the same square law for the drag of a body in a flow. It pops up in different guise from Mariotte ("... soutiennent par leur choq des poids qui sont l'un à l'autre en raison doublée de ces vitesses...") to Smeaton ("... nearly, but somewhat less than, as the square of the velocity of the wind..."). In modern terms the square law is expressed as

$$F_d = c_d \cdot \frac{1}{2}\rho A u^2,$$

where F_d is the drag of an object with the cross section area A perpendicular to the flow, ρ the density of the fluid and c_d the drag coefficient. Mariotte's data for the water flow against blades yield a drag coefficient 1.24. Newton's experiments with spheres correspond to a drag coefficient 0.5 (Flachsbart, 1932, pp. 19, 22). Of course this is a retrospective evaluation: there is no c_d coefficient in Newton's theorems, Mariotte's treatise or Smeaton's "somewhat less" square law. Yet this reconstruction shows how the drag of a body in quite different flows became accessible to measurements that arose interest still centuries later.

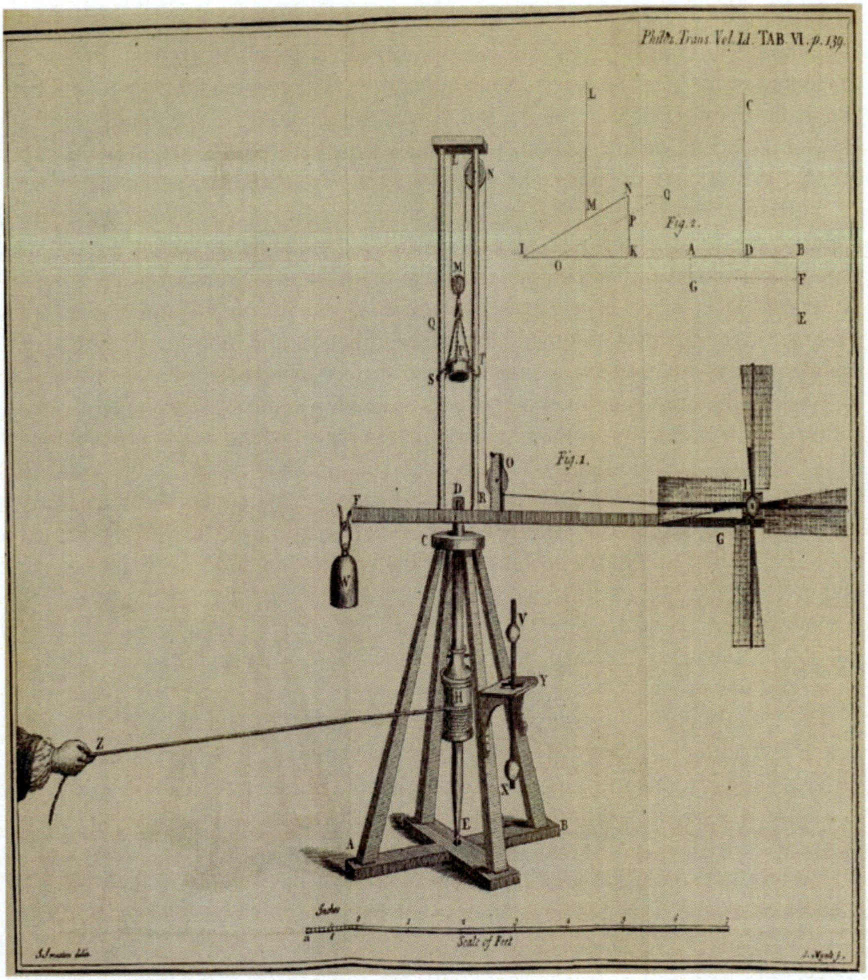

Fig. 3.2 Smeaton's whirling arm apparatus for measuring the force exerted on the sails of windmills (Smeaton, 1759, plate VI)

3.3 Ballistics

Newton's investigations on the drag of spheres became a bone of contention when compared with the experience in ballistics. In 1742 Benjamin Robins (1707–1751), a pioneer of scientific gunnery, concluded in his *New Principles of Gunnery*, "that the theory of the resistance of the air, established in slow motions by Sir Isaac Newton, and confirmed by many experiments, is altogether erroneous, when applied to the swifter motions of musket or cannon-shot". At the speed of such projectiles "the resistance is not in the duplicate proportion of the velocity of the moving body, as is usually asserted, but varies from that proportion according to the different compression of the fluid compared with the velocity" (Robins, 1742, pp. 77–78).

For slow velocities, however, Robins confirmed the square law. He measured the drag of spheres by rotating them at the end of a four feet long whirling arm (Robins, 1761). The agreement with Newton's free-fall experiments, however, was rather qualitative than quantitative. The drag measurements of a sphere (diameter of 11.5 cm) at a circular speed of 7.5 m/s led to the (retrospective) conclusion that the drag coefficient was larger than 0.55—in contrast to Newton's results. But the experiments with spheres rotating at the end of a whirling arm may not be compared to the free fall of spheres because the successive circular motion in stirred air from previous passages is different from the straight fall through quiescent air (Flachsbart, 1932, pp. 24–26).

Inspired by Robins another expert on scientific gunnery, Charles Hutton (1737–1823), extended in 1786 the whirling-arm experiments in order to determine the drag of differently shaped bodies (semi-spheres, cones, cylinders, plates) under different atmospheric conditions (air pressure controlled by barometer readings, temperature). He recorded large differences among these bodies even if they had the same cross section area (Hutton, 1812a). Once more, the retrospective evaluation revealed that Hutton's drag coefficients like those from other whirling arm measurements were systematically too large. Their virtue lies in the comparison of the resistance of bodies of different shapes to another (Flachsbart, 1932, p. 27).

The major problem of practical ballistics concerned the range of projectiles. The drag that a projectile experienced during its trajectory was crucial for this determination. Hutton regarded his whirling arm experiments only as a first step in this direction. Even more deficient he considered the contribution of theorists. As he declared in a chapter under the title "Theory and practice of gunnery, as dependent on the resistance of the air," the determination of the trajectory of a projectile moving through air was "one of the most difficult problems" to which even Newton and others contributed "no more than distant approximations". A theory that met with the expectations of practical gunnery was not in sight. All the more Hutton called for more experiments:

> The solution of this problem requires, as an indispensable datum, the perfect determination by experiment of the nature and laws of the air's resistance at different altitudes, to balls of different sizes and densities, moving with all the usual degrees of celerity. Unfortunately however, hardly any experiments of this kind have been made, excepting those which have been printed in the preceding Tracts of these volumes; which, though so extensive and varied, I cannot yet undertake to pronounce that they are fully adequate to the purpose of all the demands, either of the practical artillerist, or the speculative philosopher. (Hutton, 1812b, p. 255)

By the 19th century, therefore, drag investigations became an urgent issue on the agenda of ballistics. But gunnery was not the only practical application. Naval architecture was another.

3.4 Towing Ships

Experimental investigations on the drag of objects in a stream of water had been conducted early on, such as communicated by Mariotte and others. Whirling arm experiments like those of Robins, Smeaton and Hutton in air were also performed in water. In 1767, Jean-Charles de Borda (1733–1799), an engineer and officer in the French navy, published the results of drag measurements obtained with a whirling arm device that towed various bodies around a water basin. He noticed, for example, that the drag of bodies towed on the surface, like ships, is larger than that of fully submerged bodies (Ferreiro, 2007, pp. 160–161).

The drag of ships became also a concern for the design of navigable canals: the narrower the canal the larger the resistance that had to be overcome. In 1775, when a canal was planned that connected various waterways in Northern France through a narrow tunnel, the French government appointed a commission to carry out "theoretical and experimental research on navigation canals for the purpose of trade" (Redondi, 1985, p. 89). The members of this commission regarded "the problem of the resistance of fluids as the first and main object of our researches," as Charles Bossut, the "reporting member" of this commission, introduced the common report about this investigation (Bossut et al., 1777, p. 2). The site of their experiments was a basin at the new Paris military school. On one side of the basin a tower was erected from which boats could be towed by falling weights. By installing in the basin a wooden frame with movable floor and side walls towing tests were conducted with a variety of ship models under narrow- and shallow-water conditions (Fig. 3.3).

The experiments clearly confirmed "how important it is to give the navigable canals as much width and depth as possible". Therefore the construction of narrow canals should be avoided. "The canals under open skies deserve in general all preference over the subterranean canals."[3] Thus the commission recommended to cancel the underground canal project. Yet the plan was only postponed. In the age of Napoleon Bonaparte (1769–1821) the project was examined anew and approved more for political than scientific reasons. After a construction period of seven years, on 28 April 1810, Napoleon inaugurated the subterranean canal—just in time of the new empress, Marie-Louise's honeymoon (Redondi, 1985, p. 99).

The towing experiments at the Paris military school did not aim to distinguish between different forms of ship resistance. A first step in this direction was made by Pierre du Buat whom we have met, like Bossut, already with his investigations on pipe flow (see Sect. 1.2). Du Buat's *Principes d'Hydraulique, Vérifiés par un grand nombre d'Expériences faites par ordre du Gouvernement* dedicated a whole part to the drag of bodies, from Newton's theory of fluid friction to Bossut's recent towing experiments (Buat, 1786, Vol. II, Part III). With regard to the latter he remarked that in addition to Newton's "choc" at the front there is a wall resistance along the lateral

[3] "On voit par là combien il est essentiel de donner aux canaux de navigation, le plus de largeur et de profondeur, qu' il est possible [...] Les canaux à ciel ouvert méritent en général toute préférence sur les canaux souterrains." (Bossut et al., 1777, p. 193). For more detail see (Redondi, 1985; Ferreiro, 2007, pp. 164–167; Calero, 2008, pp. 202–214).

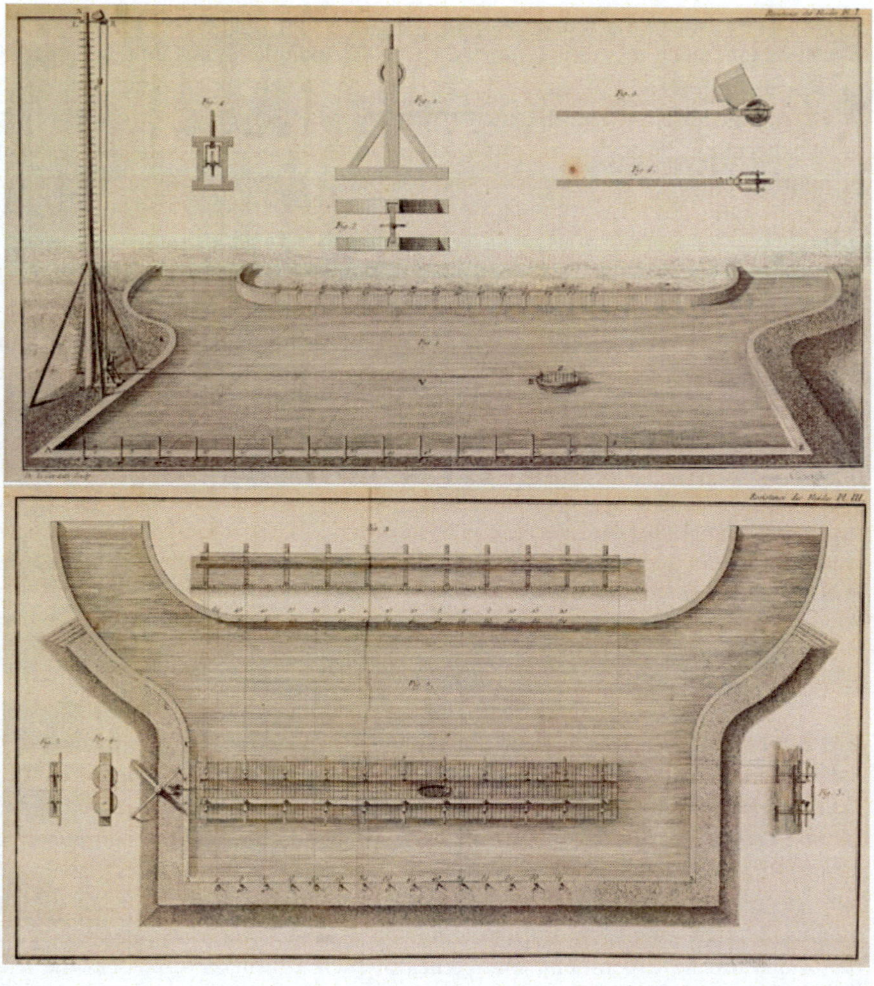

Fig. 3.3 Towing experiments for determining the drag of ships (Bossut et al., 1777, plate I and III)

ship hull and a "rétrograde" force acting from behind. The relative amount of each contribution differed from ship to ship; one estimate attributed 40–60% of the total drag to the bow shock, 20–30% to the lateral friction, and 15–25% to the retrograde force (Ferreiro, 2007, pp. 169–170).

In the1790 s, Mark Beaufoy (1764–1827), a skilled experimenter and member of the newly founded London Society for the Improvement of Naval Architecture, conducted more detailed towing trials at the London Greenland Dock (Beaufoy, 1833). They provided further evidence about the various contributions to the resistance of ships, in particular the friction along the surface of the hull. Beaufoy's dragging experiments exposed the whole complexity of the ship resistance as dependent on the shape of bow and tail, the length, and area of the hull under the surface of the water. In contrast to Newton's square law, Beaufoy concluded that "friction always increases in a much less ratio [from 1.71 to 1.82] than the duplicate ratio" [quoted in (Rouse and Ince, 1957, p. 129)].[4] In 1800 the Society for the Improvement of Naval Architecture regarded Beaufoy's "curious and instructive" towing tests with great expectation. "They clearly prove that experiments can now be made, by means of proper models, so as to ascertain the comparative advantages or disadvantages arising from the form ... of all kinds of navigable vessels" [quoted in Schaffer (2004, p. 93)].

Despite this plea for more model experiments practical naval architects had little trust in towing trials. There was no formula how to determine the resistance of a real-size vessel from the resistance measured by towing scale models through the calm water in a dock or a tank. A breakthrough in this quest was achieved only in the1870s when William Froude (1810–1879), a British civil engineer and naval architect, performed new experiments in a towing tank near Torquay (Brown, 2006). Froude assumed that ship resistance is mainly due to wave making and skin friction along the wetted surface of the hull; both obey different scaling laws (Rouse and Ince, 1957, pp. 184–185; Darrigol, 2005, Sect. 7.2.2). In modern terms, the wave pattern of model and real-size vessel is similar if the "Froude number" is the same in both cases, i. e.

$$\frac{u}{\sqrt{gl}} = \frac{U}{\sqrt{gL}},$$

where u and U is the speed, g the gravitational acceleration, l and L the length of the model and real-size boat, respectively. But this similitude only applied to wave making which originated at the bow and stern of a vessel. The investigation of skin friction concerned "the conditions of the resistance to passage through the water caused to models or ships by the friction of the water against the sides", as Froude introduced a series of "Experiments on the Surface-friction experienced by a Plane moving through water". In order to eliminate wave-making as far as possible, he performed towing experiments on "bodies of such a form as to possess the least possible displacement, and present to the line of motion the least possible sectional area, compared to the amount of wetted skin" (Froude, 1872, p. 118). He towed in his Torquay tank planks of different surface roughness and lengths edgewise at different speed—and found results which eluded a straightforward interpretation. The drag d_f for smooth and rough plate surfaces varied with different velocity exponents ϵ according to

$$d_f = c_f \gamma A u^{\epsilon},$$

[4] "With hindsight, for example, the drag coefficient of a fully submerged sphere as reconstructed from Beaufoy's measurements in the range of Reynolds numbers 1.7×10^5–1.4×10^6 was found to agree with modern measurements in the supercritical range (where the detachment of the turbulent boundary layer leads to a smaller wake than that of the laminar boundary layer at smaller Reynolds numbers). This explains the discrepancy between Newton's and Beaufoy's measurements of the sphere drag" (Flachsbart, 1932, p. 31).

where c_f is the drag coefficient for skin friction, γ the specific weight and A the wetted surface area. ϵ was in the range of 1.83 (for very smooth surfaces) to 2 (for rough surfaces). With hindsight Froude's results seemed to foreshadow to some degree the turbulent boundary layer theory as developed fifty years later (Rouse and Ince, 1957, pp. 185–186; Darrigol, 2005, p. 281).

Disregarding the unresolved nature of skin friction, the towing experiments with planks yielded "the friction of the water against the sides". Thus the wave-making contribution of a geometrically similar ship model and the surface friction of the hull as measured from corresponding plank tests could be distinguished from another and scaled up separately to give the total drag of a real-size vessel. The principle for determining the ship resistance was as follows[5]:

- Towing the model with the speed $u = U\sqrt{l/L}$ yields the total drag d of the model.
- The skin friction of the model d_f is measured from towing corresponding planks, which yields the residual drag $d_r = d - d_f$.
- The similarity of the model (mass m) and the real-size vessel (mass M) implies for the residual drag of the vessel $D_r = d_r \cdot M/m$.
- The skin friction of the vessel D_f is determined from the above skin friction formula; hence the total drag of the vessel is $D = D_f + D_r$.

Towing tank experiments as pioneered by William Froude and pursued by his son, Edmund Froude (1846–1924), after his father's death in 1879 slowly bridged the gap between science and engineering in ship hydrodynamics. The British Admiralty relocated the towing tank from Froude's private estate at Torquay to its own test station at Haslar near Portsmouth. Replication of Froude's tank met with problems due to different local circumstances and requirements, but by the beginning of the 20th century Froude's tank became the role model of a number of new test stations (Leggett, 2013; Ferreiro, 2020, Chap. 6).

The first German towing tank facilitiy was built in 1892 in Übigau near Dresden—in close collaboration with engineering professors and students of the Dresden Technical University (Gebers, 1906; Timmermann, 1975). One of them, Friedrich Gebers, took care for the modernization of the facility and used the towing tank for skin friction measurements that became the subject of his doctoral dissertation (Gebers, 1908). From these measurements Heinrich Blasius concluded that Reynolds' similarity law not only applies to pipe flow (see Sect. 1.7) but also to the skin friction of smooth plates (Blasius, 1913, p. 27). Only then the breakthrough achieved by Froude's towing experiments for the determination of ship resistance could be fully appreciated as a procedure that made use of two similarity laws: one governed by the Reynolds number (for skin friction drag), the other by the Froude number (for the residual drag mainly due to wave making).

5 Flachsbart (1932, p. 38). For a detailed contemporary account see (Durand, 1898, Chap. 1).

3.5 Drag Measurements in the Wind Tunnel

The rise of aeronautics in the early 20th century entailed a revival of drag measurements for objects moving through air. The determination of the forces on airships and airplanes called for more sophisticated means and techniques than those employed by Newton or Robins. Occasionally the old methods of investigation were pursued with improved means and new research questions, such as Otto Lilienthal's (1848–1896) rotating arm trials with plates at various angles of attack (Lilienthal, 1889, Chap. 19) or Gustave Eiffel's (1832–1923) experiments with free falling objects from the second floor of the Eiffel Tower at a height of 115 m (Eiffel, 1907). But within few years after the turn of the centuries the wind tunnel became in many countries the preferred instrument for measuring drag and lift—driven by the need to cope with these forces in "the Age of Strut-and-Wire Biplanes" (Anderson, 1997, Chap. 7).

In Germany, the first wind tunnel was installed in 1908 at the Luftschiff-Modellversuchsanstalt (Airship Model Research Institute) in Göttingen headed by Ludwig Prandtl. A year later, Prandtl was asked to present "the whole area of scientific aeronautics in lectures and exercises" at Göttingen University (Eckert, 2017, Sect. 3.4). Thus the wind tunnel served from the very beginning the dual purposes of practical model tests and scientific inquiries. In 1910 Prandtl opened a series of drag measurements in the newly founded *Zeitschrift für Flugtechnik und Motorluftschiffahrt* with "Remarks on dimensions and air drag formulae" in which he explained why he regarded it expedient

> to condense the results of the tests into informations on the coefficients that appear in those formulae; in this way the result is freed from the arbitrariness of the chosen dimensions of the specimen in a specific test, on the other side the conclusions drawn from the test are displayed in their most general way.[6]

Blasius referred to these remarks when he pleaded for a broader representation of friction coefficients as a function of the Reynolds number—not only in pipe flow (see Sect. 1.7) but also "for the air drag of wires about which there are now test results from the Model Research Institute directed by Professor Prandtl in Göttingen." Prandtl's collaborator, Otto Föppl (1885–1963), conducted these measurements so that the drag of the wires on which the models were suspended in the wind tunnel could be subtracted from the total drag of the models with the wires. For the drag of a wire (diameter d, length l) Prandtl and Föppl assumed the formula

$$W = \psi \frac{\gamma u^2}{g} dl,$$

[6] "Man tut in solchen Fällen gut, die Ergebnisse der Versuche in Angaben über die Koeffizienten zu verdichten, die in jenen Formeln vorkommen; auf solche Weise wird einerseits das Resultat von dem Zufälligen der gerade gewählten Abmessungen (o. dgl.) befreit, andererseits werden die Folgerungen aus dem Versuch in ihrer allgemeinsten Form zur Darstellung gebracht" (Prandtl, 1910a, p. 157).

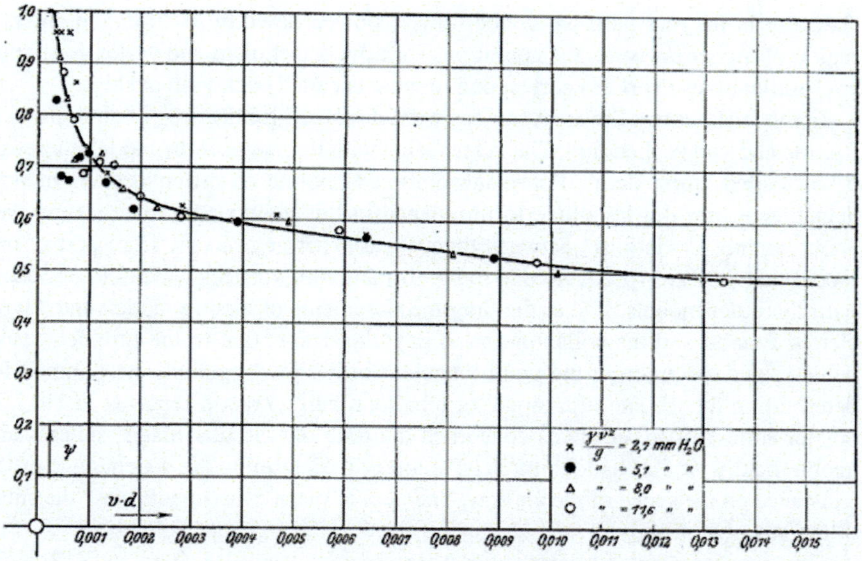

Fig. 3.4 The drag coefficient of wires measured in Prandtl's wind tunnel and plotted as a function of the product of velocity times diameter, "according to the consideration due to Reynolds" ("nach dieser Betrachtung, die von Reynolds stammt"). For large values of this product (i. e. the Reynolds number at constant viscosity) the drag coefficient approached a constant value, corresponding to the square law for the drag. For small values the drag becomes proportional to the velocity (Stokes law), i.e. the drag coefficient inversely proportional to the velocity (Foeppl, 1911, p. 85)

where W is the drag in units of lengths ("Widerstandshöhe"), ψ the dimensionless drag coefficient, γ the specific weight of air and g the gravitational acceleration. Following "the consideration due to Reynolds" Föppl displayed the drag coefficient of the wires as a function of the product ud which corresponds to the Reynolds number if the viscosity is unchanged (like in wind tunnel trials at the same atmospheric conditions); he found that the results from different measurements collapsed to data points close to a single curve (Foeppl, 1911, pp. 83–85) (Fig. 3.4).

Other measurements in Prandtl's wind tunnel concerned the drag and lift of plane and curved plates as well as models of airships. The latter became the subject of Prandtl's doctoral student Georg Fuhrmann (1883–1914). Combining theoretical methods of ideal flow theory (the method of "sources and sinks" developed by the Scottish engineer William Rankine (1820–1872) in the 19th century in an attempt to determine the optimal shape of ship hulls with minimal resistance) with wind tunnel measurements, Fuhrmann investigated the distribution of pressure and velocity around six different airship models. At the front part of the models there was close agreement between theory and experiment, but at the stern there were more or less severe differences dependent on the respective shape. "According to the theory of flow detachment by Prof. Prandtl this is completely understandable," Fuhrmann explained these discrepancies by reference to Prandtl's boundary layer concept (see Chap. 5), "because in the rear part, where the flow is slowed down by friction and enters a region of higher pressure, the conditions for flow detachment and vortex formation are fulfilled; these vortices correspond to what is called backwash at ships."[7]

In this manner two different sources were discerned for the drag of airships: skin friction and vortex shedding. The latter varied with the shape of the model; therefore it was called "form drag". Fuhrmann did not elaborate on different ways of flow detachment, nor did he allude to turbulence in this context (he merely remarked about certain "deviations" between theory and experiment in the front part of one model that they are "probably due to the formation of vortices which did not occur with the other models").[8] Yet the drag measurements of airship models must have alerted Prandtl and his collaborators to pay closer attention to the turbulent wake behind their specimen in the wind tunnel. "Further investigations are planned for determining the lee site of a model of a balloon hall", Prandtl reported in 1912. A "major series" of experiments concerned the drag of "circular plates, spheres and geometrically similar balloon models" in order to determine the dependence of the resistance on the speed of the air flow. The goal of these experiments was "the study of the law of similarity".[9]

Drag measurements for aeronautics were also conducted elsewhere. In 1912, Eiffel opened a new laboratory at Auteuil, then a Paris suburb, with two wind tunnels for air speeds up to 40 m/s (Damljanovic, 2012). In a first series of drag measurements he investigated once more the drag of spheres in order to resolve a dispute with the Göttingen laboratory. Föppl regarded the drag coefficient determined from Eiffel's measurements as too low, perhaps as a result of an erroneous evaluation of the data (Föppl, 1912). "The principal German aerodynamic laboratory, that in Göttingen, has attributed to this coefficient a value two-and-a-half times more strong", Eiffel resumed the disputed issue.[10] Eiffel's drag measurements in the new wind tunnel provided convincing evidence that the discrepancy did not result from an erroneous evaluation of his former measurements or those at Göttingen. The drag coefficient

[7] "Nach der Ablösungstheorie von Prof. Prandtl ist dies auch vollkommen erklärlich, denn an dem hinteren Ende der Körper, wo die durch die Reibung verzögerte Strömung in ein Gebiet höheren Druckes eintritt, sind die Bedingungen für die Ablösung der Strömung und die Ausbildung von Wirbeln gegeben; diese Wirbel entsprechen dem, was man bei einem Schiff als Kielwasser bezeichnet." (Fuhrmann, 1912, p. 105).

[8] "... es muss also bei diesem Modell bereits am Vorderteil eine Abweichung von der Potentialströmung stattfinden, die wohl auf die Bildung von Wirbeln zurückzuführen ist und die bei den anderen Modellen nicht auftrat." (Fuhrmann, 1912, p. 105).

[9] "An weiteren Untersuchungen ist noch die Bestimmung des Windschattens hinter einem Ballonhallenmodell und Ermittlung der Druckverteilung bei einem Gasbehältermodell zu erwähnen. 5. In der Vorbereitung begriffen sind Versuche zum Studium des Ähnlichkeitsgesetzes. Es soll an einer größeren Reihe von kreisförmigen Platten, von Kugeln und von geometrisch ähnlichen Ballonmodellen die Abhängigkeit des Widerstandes von den Abmessungen und von der Geschwindigkeit untersucht werden." (Prandtl, 1913, p. 6).

[10] "Le principal laboratoire aérodynamique allemand, celui de Göttingen, a attribué à ce coefficient une valeur deux fois et demie plus forte... Il a, du reste, publié que celui que j'avais donné moi-même était une erreur manifeste de ma part et ne pouvait être que le résultat d'une faute de calcul." (Eiffel, 1912, p. 1597).

jumped from a higher to a lower value at a critical air speed. For spheres with a diameter of 16.2 cm the phenomenon occurred at a speed of 12 m/s, for large spheres with a diameter of 33 cm between 4 and 8 m/s. The variation of the critical speed with the diameter of the sphere led Eiffel to the conclusion that

> (1) The square law for the velocity applies only for velocities higher than the critical speed [...]
>
> (2) The law of similarity does not always hold [...]
>
> (3) With regard to the total drag it is obvious, in particular for the small sphere, that there is a region where with increasing wind the drag is diminished, which seems completely paradoxical.
>
> [...]
>
> One may conclude from these anomalies that there are two regimes of flow [...] Disregarding their theoretical interest, the preceding considerations have applications in practice [...] even for airplanes.[11]

We may read in these quotes a hunch about the involvement of turbulence ("deux régimes d' écoulement"), but also a great deal of confusion. In Prandtl's laboratory Eiffel's observations launched another effort on the drag of spheres. The effect had escaped the earlier measurements because the air speed in the Göttingen wind tunnel was not high enough. By inserting a nozzle in the test section of the wind tunnel the air speed was increased from 10 to 23 m/s (Prandtl, 1913, p. 6). Eiffel's paradoxical finding was thus within reach, but Prandtl and his new collaborator in this investigation, Carl Wieselsberger (1887–1941), did not content themselves with the confirmation of the new effect. They turned the investigation of sphere drag into a fundamental study of turbulence:

> In the course of the investigations, it became apparent that the phenomena that came to light were closely related to the turbulence of the air stream in the wind tunnel. For this reason, all observations were first carried out in a homogeneous air stream, subsequently in an artificially turbulent air stream, which led to important results.[12]

Unlike Eiffel, Prandtl and Wieselsberger displayed the drag coefficient as a function of the Reynolds number ud/ν, where u is the velocity of the air stream, d the diameter of the sphere and ν is the kinematic viscosity of the air. They found that the effect occurred in Eiffel's air stream at a lower Reynolds number than in the Göttingen air stream—from which Prandtl concluded that the turbulence of the air

[11] (1) La loi du carré de la vitesse ne s' applique nettement que pour les vitesses supérieures à la vitesse critique [...] (2) La loi de similitude n'est pas toujours vraie [...] (3) Si l' on considère la résistance totale, on voit très nettement, pour la petite sphère surtout, qu' il y a une région où, quand le vent augmente, l' effort diminue, ce qui semble tout à fait paradoxical. [...] On peut se rendre compte de ses anomalies et constater qu' il existe deux régimes d' écoulement [...] Indépendamment de leur intérêt théorique, les considérations qui précèdent ont des applications dans la pratique [...] même des aeroplanes. (Eiffel, 1912, p. 1599)

[12] "Im Laufe der Untersuchungen zeigte sich, daß die zutage tretenden Erscheinungen in engster Beziehung mit der Turbulenz des Versuchsluftstromes stehen. Aus diesem Grunde wurden sämtliche Beobachtungen zuerst in einem homogenen, hierauf in einem künstlich turbulenten Luftstrom ausgeführt, die zu wichtigen Ergebnissen führten." (Wieselsberger, 1914, p. 140).

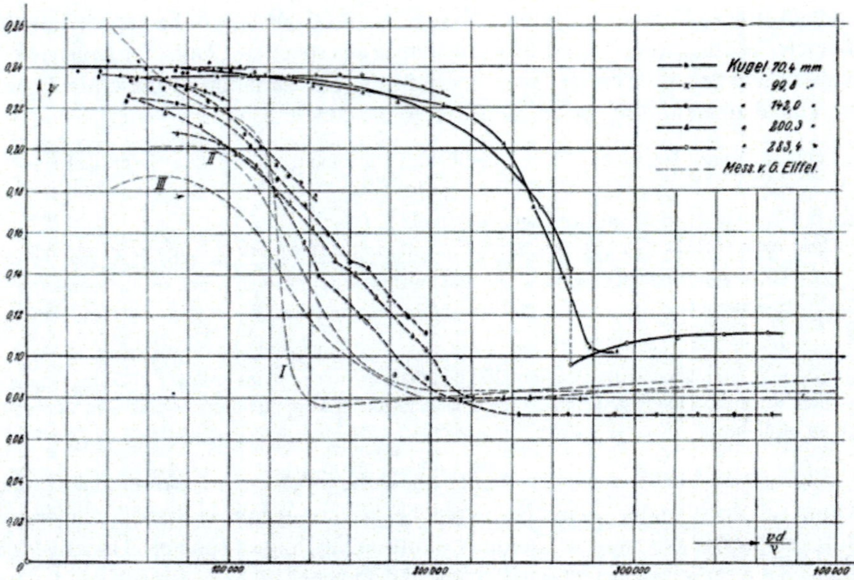

Fig. 3.5 The sphere drag coefficient as a function of the Reynolds number (Wieselsberger, 1914, p. 141)

stream plays a role in this phenomenon, and that the degree of turbulence in Eiffel's wind tunnel was higher than that in Göttingen. When they made the Göttingen air stream artificially turbulent by inserting a mesh of thin threads they could shift the critical Reynolds number to lower values (Fig. 3.5).

Prandtl also offered an explanation of this phenomenon (Prandtl, 1914): the turbulent eddies in the air stream induce transition to turbulence in the laminar boundary layer around the sphere, and the turbulent air flow in the boundary layer entrains air from the wake so that the boundary layer stays attached to the surface of the sphere longer than in the laminar case. An experimental test confirmed this explanation: when the transition to turbulence in the boundary layer was induced with a thin 'trip wire' around the sphere or a rough surface, the same phenomenon occurred. The reduced extension of the wake behind the sphere became visible with smoke added to the air stream (Fig. 3.6).

The sudden drop of the sphere drag coefficient lent itself for comparing different wind tunnels with respect to their more or less turbulent air stream:

> The obtained results are noteworthy for research laboratories where models are examined in artificial airflow. Here, the turbulent state of the flow must be taken into account, because it cannot be ruled out that similar processes also occur in other body shapes, for example in airfoils, where the separation point on the suction side usually has sufficient clearance to move. The measurement of the sphere drag offers the possibility to check the degree of

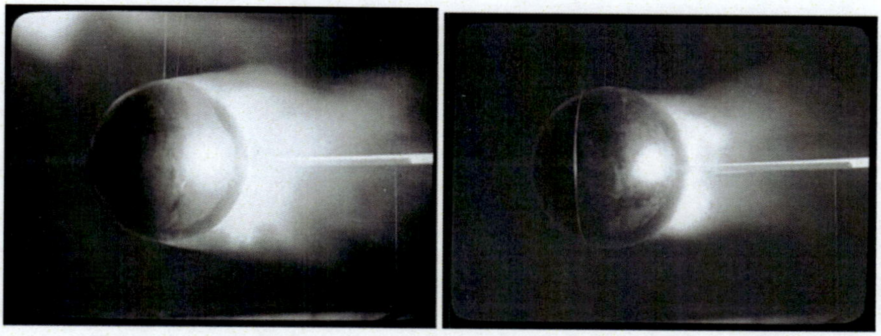

Fig. 3.6 The 'trip wire' experiment visualized the reduced extension of the wake behind the sphere (DLR archive, photographs GK-0116 and GK-0118)

turbulence of a flow by observing at which Reynolds number the transition from phase I to phase II occurs.[13]

From a theoretical perspective, Prandtl's explanation called for a theory of "the turbulence in the 'boundary layer"' (Prandtl, 1914, p. 189), a concept that had been elaborated so far only for laminar flow (see Chap. 5). Both from an applied and a more fundamental vantage point, therefore, the drag measurements in Paris and Göttingen suggested further investigations of wind tunnel turbulence.

Although turbulence was rarely addressed as explicitly as in Prandtl's laboratory in studies on the drag of bodies prior to the First World War, the wealth of drag measurements across a broad spectrum of applications—from ballistics to naval architecture and aeronautics—added to the empirical data from hydraulic engineering on pipe and channel flow that awaited theoretical evaluation. Yet there were more empirical challenges to which we turn in the following chapter before turbulence became a particular concern for closer investigations.

[13] "Die gefundenen Ergebnisse sind zunächst für Versuchsanstalten beachtenswert, in denen Modelle im künstlichen Luftstrom untersucht werden. Es ist hier vor allem auf den Turbulenzzustand der Strömung zu achten, denn es ist nicht ausgeschlossen, daß ähnliche Vorgänge auch noch bei anderen Körperformen, beispielsweise bei Tragflächen, auftreten, wo ja die Ablösungsstelle auf der Saugseite meist genügend Spiel zum Wandern hat. Die Widerstandsmessung einer Kugel bietet die Möglichkeit, den Grad der Turbulenz einer Strömung zu prüfen, indem man beobachtet, bei welcher Reynoldsschen Zahl der Umschlag von Phase I zu Phase II eintritt." (Wieselsberger, 1914, p. 145).

Chapter 4
Wind and Weather

Abstract Atmospheric turbulence has sparked scientific curiosity long ago. The causes of wind and storm became subject of numerous studies from antiquity to the 19th century. When meteorology gathered momentum as a discipline of its own right atmospheric flows were investigated from the largest to the smallest scales. "Dynamical meteorology" became the catchword for modern approaches based on thermodynamics (concerning the heat exchange of buoyant air masses) and fluid dynamics (applied to air flows over the spherical surface of the globe). In the beginning of the 20th century atmospheric turbulence was recognised as "a meteorological agency". The rise of aerology and aeronautics further contributed to gather experience with turbulence in free air by balloons and airplanes. By the 1920s the stage was set for statistical approaches on atmospheric turbulence.

4.1 Erratic Winds

Wind and weather have always called for explanation, in particular the more violent storms. The whirlwind, in Aristotle's (384–322 BC) view, "is due to the resistance which generates the eddy, and it consists in the spiral which descends to the earth and drags with it the cloud which it cannot shake off." This passage is contained in Book III, Part 1 of Aristotle's treatise *Meteorologica*[1]—a term that included for the ancient Greeks far more than the modern designation of meteorology. Three centuries later, the city of Athens dedicated to the wind a large octagonal marble tower topped with a weather vane. Whether it may be regarded as "the world's first meteorological station"[2] or rather a monument for the eight wind gods depicted in the frieze surrounding the tower—it certainly testifies to the great importance attributed

[1] Translated by E. W. Webster. In: The Internet Classics Archive. Available online at http://classics.mit.edu//Aristotle/meteorology.html. See also (Solari, 2019, Sect. 2.2).

[2] https://en.wikipedia.org/wiki/Tower_of_the_Winds. In the absence of evidence for meteorological measurements it seems more plausible that "this temple was probably a place where the devout could offer prayers and gifts in view of obtaining the wind and weather most desired for agricultural and nautical purposes" (Frisinger, 1977, p. 119).

M. Eckert, *Turbulence—an Odyssey*, History of Physics,
https://doi.org/10.1007/978-3-030-91459-2_4

to the wind in ancient Greece. The Romans, too, paid tribute to the wind. Seneca (ca. 0–65) merged Aristotle's views with Democrit's (460–370 BC) atomistic philosophy into a concept that compared atmospheric motions to the behaviour of a crowd of people:

> Democritus avers that when there are many particles, which he calls atoms, in a small empty space, wind is the outcome. But, on the contrary, when the space is large and the particles few, there is a still peaceful condition of the atmosphere. To illustrate: in the market square or in a side street as long as there is a sprinkling of people there is no disturbance as one walks along it; but when a crowd meets in a narrow space, then they jostle against each other, and quarrelling arises. Similarly in this space which surrounds our earth; when many bodies have crowded a very small portion, it is unavoidable that they should jostle one another and be driven back and forward, and be intertwined and squeezed. Hence results wind; the particles that were struggling have had to give way, and after being tossed about and remaining in suspense for a long time they at length lean their weight toward one side. But when a few bodies occupy a large roomy place, they can neither ram each other nor be jostled by one another. (Clarke and Geikie, 1910, pp. 194–195)

We should not read a prescient hunch of turbulence between the lines of this quote. It merely illustrates how atmospheric motions entered ancient natural philosophy—almost two-thousand years before "meteorlogica" became narrowed into the scientific discipline designated by the same name. The dawn of scientific meteorology arose with the New Science proclaimed by Galileo Galilei, driven mainly by the development of instruments and facilities for atmospheric measurements (Frisinger, 1977). In this process the erratic winds were a permanent challenge. In 1663 Robert Hooke (1635–1702), for example, suggested to the Royal Society a "Method for Making a History of the Weather" for which he regarded it "requisite to observe, 1. The Strength and Quarter of the Winds, and to register the Changes as often as they happen" (Sprat, 1667, p. 173). Among other meteorological instruments, Hooke designed an anemometer which measured the strength of the wind by the deflection of a plate. It was in use until it was replaced in the 19th century by pressure tubes and rotational devices. "For many, Robert Hooke is the Father of Modern Meteorology" (Frisinger, 1977, p. 102).

Edmond Halley (1656–1742), better known for predicting from Newton's laws the return of a comet that now bears his name, carried the investigation of winds one step further by looking for more systematic patterns in the distribution of wind over the whole earth. In 1687 he published "An Historical Account of the Trade Winds, and Monsoons, Observable in the Seas between and Near the Tropicks, with an Attempt to Assign the Phisical Cause of the Said Winds" where he reviewed wind observations made at sea voyages on a global scale. As the cause of the winds he discerned "the Action of the Sun Beams upon the Air and Water" which results in variations of air density: "the Air which is less rarified or expanded by heat, and consequently more ponderous, must have a Motion towards those parts thereof, which are more rarified, and less ponderous, to bring it to an Aequilibrium" (Halley, 1687, p. 165). The distribution of heat over the surface of the earth, therefore, entailed thermal convections that resulted in a global atmospheric circulation. Thus Halley entered the history of meteorology as "a pioneer in the determination of the forces acting upon the atmosphere. For his significant contributions, he is often called the

Father of Dynamic Meteorology" (Frisinger, 1977, p. 125). The global perspective of atmospheric flows received further support in 1735 when George Hadley (1685–1768) published a short paper "Concerning the Cause of the General Trade-Winds". He argued

> that the Air, as it moves from the Tropicks towards the Equator, having a less Velocity than the Parts of the Earth it arrives at, will have a relative Motion contrary to that of the diurnial Motion of the Earth in those Parts [...] And thus the Air will continue to circulate, and gain and lose Velocity by Turns from the Surface of the Earth or Sea, as it approaches to, or recedes from the Equator. (Hadley, 1735, pp. 60–62)

While the experience of seamen sailing in regular trade-winds suggested theories of large-scale atmospheric circulations, the irregular local winds observed by weather vanes and anemometers on land escaped a global explanation. Already in the beginning of what became the longest series of weather recordings, the "Hohenpeißenberger Reihe", the observer reviewed in 1792 the first ten-year-record almost with dispair. It appeared impossible to him "in view of so many and manifold phenomena that occur in the atmosphere to ever find an exit from this maze".[3]

Yet there was no lack of studies that attempted to discern general causes of the origins of wind. Like in the age of Aristotle, the most violent storms and whirlwinds—by now called cyclones—became subject of most sophisticated inquiries. Heinrich Wilhelm Dove (1803–1879), for example, since 1849 director of the Prussian Meteorological Institute (Preußisches Meteorologisches Institut), published in 1857 a treatise on *On the Law of Storms* with numerous references to recent storm studies such as "An attempt to develop the law of storms by means of facts arranged according to place and time and hence to point-out a cause for the variable winds" (1838), "The philosophy of storms" (1841), "On Whirlwind storms with replies to the objections and strictures of Dr. Hare" (1842), "The progress of the development of the law of storms and of the variable winds" (1849), to quote only the titles of some studies that evoked debates on this issue by the mid of the 19th century (Dove, 1857). In his own treatise, Dove discerned the changes of the air pressure due to opposing streams of air as a main indicator of storms. The combined experience from two centuries pointed to the importance of barometric measurements for the prediction of storms, "so that even today as a rule the designation 'very stormy' concludes the weather scale of our barometer", this is how Dove reviewed practical evidence about the fall of barometer with the incidence of storms since Otto von Guericke's (1602–1686) experiments on air pressure. He filled page after page with records of storms all over the earth, and he presented his observations and conclusions in a popular language: "The weather vane may jump into the opposite direction if steady winds push against another, fence with each other, as the seaman says."[4]

---

[3] "Freylich scheint es bey dem ersten Anblicke so vieler und mannigfaltiger Erscheinungen, die sich in der Atmosphäre ereignen, unmöglich zu seyn, jemals in diesem Labyrinth einen Ausgang zu finden". Quoted in Wege (1996, p. 91).

[4] "Überhaupt hat sich die Bemerkung des Otto von Guerike im Laufe der seitdem verflossenen zwei Jahrhunderte so vielfach bestätigt, dass die Wetterskalen unserer Barometer heute noch in der Regel mit der Bezeichnung 'sehr stürmisch' schließen. [...] Die Windfahne kann aus einer Richtung

4.2 The Physics of Atmospheric Flows

Hydrodynamics and the rise of thermodynamics by mid-19th century turned meteorology into a physical science (Garber, 1976). Kinematic considerations like those of Dove and others were considered as deficient with respect to physical principles. "It can hardly be disputed that Dove's completely non-physical theories retarded the progress of meteorology for a long time", the Austrian meteorologist Julius von Hann (1839–1921) remarked in a study on the Foehn, a dry warm wind in the Alps which had been subject of controversial debate among meteorologists (Kutzbach, 1979, p. 61). Hann explained the Foehn as the result of an adiabatic change of temperature due to compression and expansion of atmospheric flow over a mountain—an explanation based on the physics of heat as developed by William Thomson (Lord Kelvin) (1824–1907), Hermann von Helmholtz (1821–1894) and others.

For physicists like Helmholtz, meteorology became a kind of proving ground for illustrating physical principles. Even before Hann's explanation of the Foehn, Helmholtz used this phenomenon as an example for adiabatic heat exchange:

> In a mass of air that expands, a part of its supply of heat disappears; it becomes cooler if it cannot absorb new heat from outside. Conversely, by compressing the air again, the same amount of heat that had disappeared as a result of the expansion is produced again. If, for example, southerly winds drive the warm air of the Mediterranean towards the north and force it to rise to the height of the great mountain wall of the Alps, where the air expands by about half its volume, according to the lower air pressure indicated by the barometer, it cools down considerably [...] and at the same time deposits a greater proportion of its moisture as rain or snow. When the same air subsequently descends into valleys and plains on the northern side of the mountains as a foehn wind, it is compressed and warms up again.[5]

Besides thermal physics novel concepts from hydrodynamics found application in late 19th century meteorology. One of Helmholtz's pet research subjects concerned vortex motion (Darrigol, 1998). He defined notions like vorticity, vortex lines, vortex filaments, vortex tubes, vortex sheets and derived theorems of vortex motion in ideal fluids. Closely related to these concepts was the picture of discontinuity surfaces which could be considered as the limiting case of vortex sheets. Such surfaces are

---

unmittelbar in die entgegengesetzte überspringen, wenn die Winde einander stauen, mit einander fechten, wie der Seemann sagt" (Dove, 1857, pp. 11, 62–63).

[5] "In einer Luftmasse nämlich, welche sich ausdehnt, verschwindet ein Theil ihres Wärmevoraths, sie wird kühler, wenn sie nicht neue Wärme von aussen aufnehmen kann. Umgekehrt wird durch erneutes Zusammendrücken der Luft dieselbe Wärmemenge wieder erzeugt, welche durch die Ausdehnung verschwunden war. Wenn also zum Beispiel Südwinde die warme Luft des Mittelmeers nach Norden treiben, und sie zwingen zur Höhe des grossen Gebirgswalls der Alpen hinauf zusteigen, wo sich die Luft, entsprechend dem geringeren durch das Barometer angezeigten Luftdrucke, etwa um die Hälfte ihres Volumens ausdehnt, so kühlt sie sich dabei auch sehr beträchtlich ab – für eine mittlere Höhe des Gebirges von 11000 Fuss um 16 bis 25° R.[éaumur]. je nachdem sie feucht oder trocken ist – und dabei setzt sie auch gleichzeitig den grösseren Theil ihrer Feuchtigkeit als Regen oder Schnee ab. Kommt dieselbe Luft nachher auf der Nordseite des Gebirges als Föhnwind wieder in Thäler und Ebenen hinab, so wird sie wieder verdichtet und erwärmt sich auch wieder." Helmholtz in a lecture on "Eis und Gletscher", held in February 1865 and printed in (Helmholtz, 1865, p. 97).

found between fluid layers moving at different velocity—and prone to instability: the slightest waviness would increase and turn the surface into a sequence of breaking waves and whirls. In 1886 Helmholtz remarked in a talk on the formation of clouds and thunderstorms, that he had observed cloud patterns of this kind when he was hiking in the Swiss Alps. The whirls which are formed at surfaces of discontinuity produce a mixing of differently moving air layers, he argued two years later in a paper "On atmospheric motions". Mixing caused by vortex formation at discontinuity surfaces (Fig. 4.1), rather than the molecular viscosity of air, brake down the upper large-scale atmospheric circulation to the winds and storms observed on the ground.[6]

By the turn of the centuries a growing number of meteorologists adopted the views of Helmholtz and others who regarded weather phenomena from the perspective of thermodynamics and the physics of atmospheric flows. "Dynamical meteorology" became the catchword for modern approaches based on the heat exchanges of air masses and the fluid dynamics of air flows. An untiring protagonist for making the physics of the atmosphere the core of meteorology was William Napier Shaw (1854–1945), a physicist from the Cavendish Laboratory at Cambridge who became the leading representative of meteorology in England. Another hotbed was the Central Institute of Meteorology and Geodynamics (Zentralanstalt für Meteorologie und Geodynamik, ZAMG) in Vienna (Hammerl et al., 2001). In Germany, an outstanding pioneer of dynamical meteorology was Vilhelm Bjerknes (1862–1951), a former assistant of Heinrich Hertz (1857–1894) and since 1912 professor of geophysics at the university of Leipzig (Friedman, 1989). Later he became famous as head of the "Bergen school" in Norway for the development of the so-called "Polar-front theory of atmospheric circulation"; but he had already in 1898 published a groundbreaking paper "On a Fundamental Theorem of Hydrodynamics and Its Applications Particularly to the Mechanics of the Atmosphere and the World's Oceans" that extended Helmholtz's and Kelvin's theorems of vortical motion for geophysical applications (Thorpe et al., 2003; Kutzbach, 1979, Sect. 6.3). Bjerknes' circulation theorem enabled, for example, the prediction of cyclones from measurements of varying pressure and density distributions. But Bjerknes did not aim primarily at an application to one or another geophysical phenomenon. First and foremost he was eager to put meteorology on the solid grounds of physics. His theorem provided a "rational principle around which one can group the facts found through observations", to quote Bjerknes's own judgment about the importance of his circulation theorem [quoted in Kutzbach (1979, p. 163)].

Among the German meteorologists it was in particular Hugo Hergesell (1859–1938), director of the meteorological service in Strassburg, who made himself a name as a sedulous missionary of aerology as the new science of the entire atmosphere. Among meteorologist he was famous by his nickname "Hin- und Her-Gesell" ("Back-

[6] "Ich ziehe aus den dargelegten Erwägungen das Ergebniss, dass die hauptsächlichste Hemmung der Circulation unserer Atmosphäre, welche verhindert, dass dieselbe nicht ausserordentlich viel heftigere Winde erregt, als es thatsächlich der Fall ist, nicht so wohl in der Reibung an der Erdoberfläche als in der Vermischung verschieden bewegter Luftschichten durch Wirbel gegeben ist, die durch Aufrollung von Discontinuitätsflächen entstehen." (Helmholtz, 1888, p. 308); see also (Darrigol, 1998, pp. 40–41).

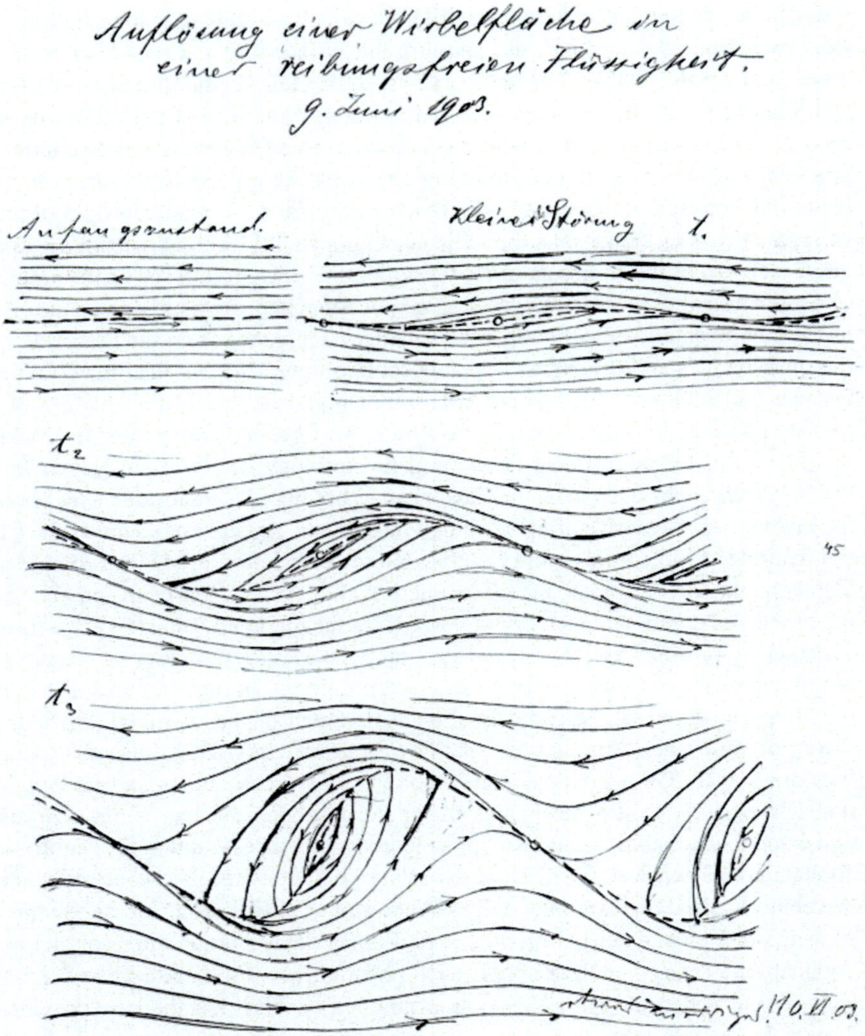

Fig. 4.1 Formation of eddies by the dissolution of a discontinuity surface, as visualized by Ludwig Prandtl (SUB Göttingen, Cod. Ms. L. Prandtl, Nr. 14, p. 45)

and Forth Fellow"). Bjerknes, himself performing as an untiring "missionary among the heathens" among more conventional meteorologists, persuaded Hergesell in 1912 to adopt notions from physics like gravitational potential instead of height as more rational units in aerology. Both agreed that aerology should be founded on the solid pillars of physics—not the least for practical concerns such as the sudden appearance of gusts which caused accidents among balloonist and pilots (Friedman, 1989, pp. 77–78).

The Viennese meteorologists, too, attempted to make meteorology an exact science. Max Margules (1856–1920), a theoretical physicist employed since 1882 at the ZAMG under the directorship of Hann, developed a theory of cyclones from physical principles. In a paper published in 1903 "On the energy of storms" Margules focused on the origin for the kinetic energy of a cyclone. "The kinetic energy of a mass of air is derived from its internal energy and from the work done by the force of gravity", he invoked the theorem of energy conservation for the genesis of storms. "So far as I can see the source of storms is to be sought only in the potential energy of position" [quoted in Garber (1976, p. 62)]. In his cyclone theory he also extended Helmholtz's concept of surfaces of discontinuity so that it involved air masses at different temperatures. "His work in part anticipated and supported later investigations on frontal slopes by meteorologists in Bergen around 1920," this is how Margules entered the history of modern meteorology (Kutzbach, 1979, p. 199).

Another pioneer for the modernization of meteorology was Felix Exner (1876–1930), a colleague of Margules at the ZAMG. In 1912 he gave the modern views a name with an encyclopedia article titled "Dynamische Meteorologie" (Exner, 1912). A few years before he had published "Fundamentals of a Theory of Synoptic Changes of Air Pressure" ("Grundzüge einer Theorie der synoptischen Luftdruckveränderungen") where he used a simplified set of differential equations for predicting from barometric measurements the surface pressure of the air over an extended region for several hours. Furthermore, by qualifying his prognosis with a certain percentage of likelihood, he introduced "a probabilistic style of reasoning that has since become ubiquitous in meteorology" (Coen, 2007, p. 285). Like for other modern meteorologists the genesis of cyclones became a favourite target of Exner's theoretical efforts. In competition with the polar front theory of the Bergen school which assumed a wavy boundary between cold and warm air masses all around the globe, Exner allowed for interruptions through which polar air could break through. Whereas the Bergen school discerned convolutions of the polar front as the cause of cyclones, Exner emphasized the instability of the front that would make turbulence an important ingredient for a theory of cyclogenesis (Coen, 2007, p. 290; Volkert, 1999).

4.3 How Turbulence Became a Meteorological Agency

Thus the "Vienna school" of dynamical meteorology prepared the ground for a concept which became widely known in the 1920s under the label "Austausch" (exchange). It was elaborated by Exner's close colleague (and future successor as director of the ZAMG) Wilhelm Schmidt (1883–1936) in a widely read book under the title *The Mass-exchange in Free Air and Related Phenomena* (Schmidt, 1925). A book review addressed to the meteorological community draw from Schmidt's account the lesson that progress in meteorology requires "a full knowledge of the conditions and consequences of irregular flow in gases and liquids".[7]

[7] Quarterly Journal of the Royal Meteorological Society, 53, issue 221, 1927, pp. 93–94.

Schmidt focussed on "irregular flow" in meteorology already in May 1917, when he presented to the Austrian academy of science a memoir on "The Mass-exchange due to Irregular Flow in Free Air and its Consequences" ("Der Massenaustausch bei der ungeordneten Strömung in freier Luft und seine Folgen"). The focus on turbulence is clear from the very beginning, although the setting in quotation marks indicates the novelty of this designation in meteorology. In a footnote Schmidt remarked that the term "turbulence" so far referred to experiments in the laboratory:

> The behaviour of flowing liquids had led to the assumption of a special state of irregular, "turbulent" motion. [...] It is now natural that with the assumption of irregular motion all those physical laws and constants were no longer applicable in which the migration of the individual molecules plays a decisive role: these are those for heat conduction, diffusion, internal friction, which result from the kinetic theory of gases. [...] In the following work, a quantity, called exchange, is derived which is decisive for a number of such processes. It allows the transfer of heat, motion, masses etc. between the different layers in free air to be calculated according to known formulae, such as those which apply to ordinary heat conduction, it combines processes which have hitherto only been considered separately, if at all, and thus supports the observations on one meteorological element by those on the other. Only in this way does it become clear how important the irregular motion is for mixing in the atmosphere.[8]

The rational of Schmidt's concept was to derive from the mass exchange not only the transfer of momentum between different layers of air but also the diffusion of heat, water vapour or dust in the atmosphere. The emphasis was on the irregular or disordered manner of this mixing. Schmidt's exchange-concept was an attempt to cope with the role of turbulence in atmospheric processes. He regarded the exchange of mass between adjacent layers of air and the entailed mixing as a result of eddying motion rather than molecular diffusion. The coefficient used to account for the latter process, therefore, had to be replaced by another quantity which Schmidt named Austausch (exchange) and designated by the letter A. This exchange-constant replaced in formulae for mixing processes the molecular diffusion constants and friction coefficients, but unlike the latter Schmidt's A was no constant but depended on local conditions. Empirical support for Schmidt's concept came from wind speed measurements conducted for many years at different heights from the Eiffel tower which showed a systematic increase of mean wind speeds with height (Akerblom, 1908). Other measurements were obtained from kites and pilot balloons. Schmidt derived

[8] "Das Verhalten strömender Flüssigkeiten hatte zur Annahme eines besonderen Zustandes ungeordneter, 'turbulenter' Bewegung geführt. [...] Es ist nun natürlich, daß mit der Annahme ungeordneter Bewegung alle jene physikalischen Gesetze und Konstanten nicht mehr anwendbar waren, bei denen das Wandern der einzelnen Moleküle eine maßgebende Rolle spielt: das sind die für Wärmeleitung, Diffusion, innere Reibung, die sich aus der kinetischen Gastheorie ergeben. Ihnen entsprechen hier andere, allgemeinere und weniger bestimmte Gesetze. [...] In der folgenden Arbeit wird nun eine Größe, Austausch genannt, abgeleitet, die für eine Reihe solcher Vorgänge maßgebend ist. Sie gestattet, Übergang von Wärme, Bewegungsgröße, Massen usw. zwischen den verschiedenen Schichten in freier Luft nach bekannten Formeln zu rechnen, wie sie etwa für die gewöhnliche Wärmeleitung gelten, sie faßt Vorgänge zusammen, die bisher, wenn überhaupt, nur getrennt betrachtet wurden, und stützt so die Beobachtungen an einem meteorologischen Element durch solche am anderen. Auf diesem Wege wird erst klar, wie bedeutsam die ungeordnete Bewegung für die Umsätze in der Atmosphäre ist." (Schmidt, 1917, pp. 757–758).

from such data a value of A that gradually increased up to a height of 500 m from where it decreased again to a somewhat lower value (Schmidt, 1917, p. 771). Thus A reflected different degrees of turbulence at different heights above ground.

Schmidt's attempt to quantify atmospheric turbulence was not unprecedented. In May 1914, Napier Shaw had presented to the Royal Society a paper by Geoffrey Ingram Taylor (1886–1975) on "Eddy Motion in the Atmosphere" which was published in 1915 in the first issue of the *Philosophical Transactions of the Royal Society* of this year—but apparently did not reach Austria due to the outbreak of World War I. Taylor, then a young physicist employed at Cambridge University as "Reader in Dynamical Meteorology", started with the propagation of heat upwards or downwards through the atmosphere. "The ordinary conductivity of heat by molecular agitation is so small that no sensible error will be introduced by leaving it out of the calculations," he dismissed molecular diffusion for this transport in favour of a propagation by means of eddies. The latter yielded an equation that resembled the heat conduction equation—with the coefficient of thermal conductivity replaced by a quantity that Taylor called "eddy conductivity" (Taylor, 1915, p. 4). After the transport of heat by eddying motion he turned to the transport of momentum. "It seems natural to suppose that eddies will transfer not only the heat and water vapour, but also the momentum of the layer in which they originated to the layer with which they mix. In this way there will be an interchange of momentum between the different layers." Taylor named the quantity that characterized this interchange "eddy viscosity" (Taylor, 1915, p. 13). His reasoning, by and large, agreed with Schmidt's exchange-concept, but where Schmidt was content with plausibility arguments and rough numerical estimates, Taylor presented his views in the form of a mathematical framework and compared his theory with measurements that he had made during an expedition on board of an ice-scout ship in the North Atlantic (Batchelor, 1996, Chap. 5). Although he did not (yet) use the word "turbulence", he addressed the irregular motion of atmospheric flows as a subject that is accessible to mathematical theory and comparison with meteorological data. "These calculations are very rough," he closed his article, "but they do show at any rate, that actual observations of eddy motion do not involve anything that is contrary to the assumptions on which the theory contained in this paper is based." (Taylor, 1915, p. 22).

Another approach at atmospheric turbulence originated at a remote meteorological station in Scotland. "Serious attention to the problem was begun in 1913 at Eskdalemuir Observatory with the permission and encouragement of Sir Napier Shaw, then Director of the Meteorological Office," this is how Lewis Fry Richardson (1881–1953) recalled the beginnings of his meteorological investigations a few years later. "The problem" was an attempt to solve the differential equations that describe the flow of air and the local state of the atmosphere approximately by transforming them into a set of difference equations that could be solved at discrete points of a three-dimensional grid. In the solitude as superintendent of this observatory, Richardson turned "the fantasy" into a serious research program. The war interrupted these investigations. In 1920, when he was employed at the Meteorological Office in Benson, Richardson resumed his efforts. "The whole work has been thoroughly revised in 1920, 1921," he summarized the development that led to his book on *Weather Prediction by Numerical Process* (Richardson, 1922, pp. viii–ix).

The goal of numerical weather forecast (Nebeker, 1995, Chap. 6; Lynch, 2006) must have sparked Richardson's attention for atmospheric turbulence from the very beginning because irregular flow seemed to elude a computational scheme. It is not astonishing, therefore, that he closely studied the work of Taylor and Schmidt as soon as he became aware of it after the war. "The Supply of Energy from and to Atmospheric Eddies," as he titled a first paper on this subject in 1920, served him as a starting point for this investigation. The eddies are "acting as thermodynamic engines in a gravitating atmosphere" and from the energy supplied or consumed in this process one could derive a criterion for turbulence in atmospheric flow, analogous to Reynolds' criterion for turbulence in pipe flow. The analysis amounted to a comparison of two contributions to the kinetic energy of the eddies: one due to the buoyancy that carries eddies upwards or downwards according to the temperature gradient; the other due to the acceleration of eddies between different wind speeds. If the latter exceeded the former, the flow would become more turbulent. "To take a special application of it," Richardson illustrated this criterion by an example, "suppose that the large piece of atmosphere, throughout which the integral is taken, is initially at rest. Suppose next that an approaching depression sets it in motion. Will the winds be turbulent or not?" The formula which he had derived for the respective energy contributions answered this question (Richardson, 1920b, p. 364). Many years later, the fraction of these contributions became called Richardson number.

The paper on "The Supply of Energy from and to Atmospheric Eddies" marked the beginning of Richardson's ground-breaking publications on atmospheric turbulence (Hunt, 1998; Benzi, 2011). In the same context, Richardson attempted to discern an appropriate physical quantity for turbulence. "ξ might be named 'the turbulivity'," he labelled one such quantity. "It is suited to the free atmosphere" but "has no name derivable from indoor physics" (Richardson, 1920a, p. 2). Two years later he dedicated an entire chapter to turbulence in his *Weather Prediction by Numerical Process*. The numerous references to Schmidt's and Taylor's papers therein also exposed the common features of their work. What Schmidt called "Austausch" corresponded closely to Taylor's and Richardson's notions of "eddy viscosity". In this chapter we also find the famous rime about the cascade-like process that propagates turbulence from the largest scales of cyclones down to molecular dimensions (Richardson, 1922, p. 66):

> We realize thus that: big whirls have little whirls that feed on their velocity, and little whirls have lesser whirls and so on to viscosity—in the molecular sense.

This verse became a leitmotiv in the history of turbulence. It was used for the first time when Napier Shaw reviewed in *Nature* an article of another dynamical meteorologist from the Vienna School, Albert Defant (1884–1994), in which Schmidt's "Austausch" concept was extended from a vertical mass exchange in the atmosphere to a large-scale turbulent horizontal mixing. This study illustrated in Napier Shaw's view to what extent turbulence must be perceived as "a meteorological agency". Its study had begun before the war and "developed independently on either side of the front during the war", Napier Shaw remarked with reference to the investigations of Exner, Schmidt, Taylor, Richardson and few others. Shaw's distinction of pioneering work "on either side of the front" also reflects the mentality of scientists shortly after the war when they struggled to overcome their wartime division into hostile camps (Shaw, 1922, p. 469):

> Both groups in different ways have come to recognise that transference by irregular turbulent motion is a principle of very general application, and that turbulence may range step by step, perhaps discontinuously, from the ultra-microscopic scale of molecular turbulence which we measure as viscosity to the gigantic scale of the eddy motion of the great cyclonic depressions or even of the general circulation of the whole atmosphere. L. F. Richardson has expressed the generality in a manner which easily sticks in the memory:—
> Big whirls have lesser whirls that prey on their velocity;
> They again have smaller whirls, and so on to viscosity.

4.4 Experiencing Turbulence in World War I

Turbulence not only became a meteorological agency from a theoretical vantage point. "Airmen are very familiar with the increased bumpiness of the wind caused by sun shining on the ground below them", Richardson referred to the practical aspects of atmospheric turbulence. "The upward currents of large cumuli give rise to much turbulence within, below, and around the clouds, and the structure of the clouds is often very complex." This was one such observation which Richardson quoted in the context of his big-whirls-little-whirls-rime (Richardson, 1922, p. 66). The "airman" to whom Richardson referred in this case was Charles Kenneth Mackinnon Douglas (1893–1982), a Lieutenant of the Royal Flying Corps, who had reported extensively about his "Weather Observations from an Aeroplane" in 1916 in the *Journal of the Scottish Meteorological Society* (Douglas, 1916). These observations were made in Northern France during spring and summer of 1916 up to heights of about 10,000 ft. "No one who reads this paper," a reviewer praised Douglas, "can fail to realise how much our knowledge of the conditions in the upper air may be increased by the observations of a capable observer flying daily up to those heights which are easily within the compass of an aeroplane."[9]

Taylor's views of atmospheric turbulence, too, were accompanied by observations made in the context of aeronautics. In the First World War he served as a kind of general problem solver at the Royal Aircraft Factory in Farnborough. "I felt that I might be able to do a better job if I were taught to fly myself," he later recalled (Batchelor, 1996, Chap. 6). In July 1915 he graduated as a pilot. A year later he conducted measurements of the pressure distribution along a wing section mounted on the wing of his airplane (Batchelor, 1996, p. 71). During the final two years of the war his reports to the Advisory Committee for Aeronautics show that he was mainly concerned with investigations on atmospheric turbulence. At the request of the Meteorological Office he conducted a set of observations on the "Variation of

---

[9] Review in *Quarterly Journal of the Royal Meteorological Society*, 43:182, 1917, pp. 227–229.

Wind Velocity Close to the Ground"—motivated by the use of poison gas at the front. Knowledge of turbulent surface winds should provide "information for use of officers responsible for discharging the gas during a gas attack" (Taylor, 1916). Other studies dealt with "The Formation of Fog and Mist" (Taylor, 1917c), "Observations and Speculations on the Nature of Turbulent Motion" (Taylor, 1917a), "Phenomena Connected with Turbulence in the Lower Atmosphere" (Taylor, 1917b) and "On the Dissipation of Eddies" (Taylor, 1918).

The rise of meteorology in World War I was conspicuous. An observer from Harvard University noticed already in 1915 the "new and curious emphasis upon the importance of weather controls in modern warfare". By then the USA was not yet participating in the war, but the growing importance of meteorology on the battlefields did not fail to raise concern. The use of poison gas provided an outstanding example. During a gas attack of the Germans the wind "suddenly changed, blowing these same gases back to the German lines, and forcing the men to flee for their lives" (Ward, 1915, p. 317). The *Journal of Geography* paid tribute to the new emphasis on meteorology with a series of articles under the title "The Weather Factor in the Great War" (Bates and Fuller, 1986, p. 17; Rollins, 2017, p. 10). In the fourth paper of this series the author noticed that the Germans had learned from their previous ignorance about the meteorological conditions during a gas attack. "In regard to the liberation of asphyxiating gas, whose safe and effective use is impossible unless the wind is in the right direction and of the proper velocity, the Germans have been sending up kites in order that they may determine the probable wind direction beforehand, and govern themselves accordingly" (Ward, 1916, p. 384). Such observations contributed to establish meteorological programs in American universities after the war—although "it would prove to be a very slow process" compared to Europe (Harper, 2006, p. 184).

Another lesson for postwar meteorology concerned its close ties to aviation. "Meteorology and Aviation" was the title of a paper published in 1919 by Ernest Gold (1881–1976), a scientist recruited to the Meteorological Office in World War I (Crewe, 2009). Under the same headline, C. K. M. Douglas from the Royal Flying Corps presented to the readers of *Nature* first-hand observations of the turbulent mixing of air masses, illustrated with cloud photographs at high altitude. "At greater heights eddy motion may be developed throughout a layer 1000 ft. or 2000 ft. thick when the lapse-rate of temperature (vertical temperature gradient) equals the adiabatic rate owing to the intrusion of a cold layer, and if the humidity is high enough clouds may form at the top of the turbulent layer. Sheets of rippled 'alto-cumulus' clouds often develop in this way" (Douglas, 1919, p. 474). This was just one example how observations from light airplanes in World War I added atmospheric turbulence to the agenda of postwar meteorology. "With some oversight from the Advisory Committee for Aeronautics, the understanding of the aviator's environment improved dramatically," this is how the historian of the British Meteorological Office briefly summarized the legacy of this war for meteorology (Crewe, 2009).

The same experiences were made by the other belligerent powers in World War I. In Germany the rise of aviation had suggested already before the war the formation of military weather services for aerial operations. With the outbreak of war German

Fig. 4.2 Preparation of wind measurements at a German meteorological field station in occupied Belgium in the beginning of WW I (courtesy of Cornelia Kemp)

meteorologists re-oriented their work along military goals. Data from field stations (Fig. 4.2) were kept secret (Oestreich, 2008). When some of the meteorologists' war work was published after the war it became obvious to what extent turbulence had become a particular concern. "Observations about the turbulence of wind, particularly in the free atmosphere", such was titled for example a report from a naval weather station in occupied Belgium (Peppler, 1922, pp. 19–32). The Germans also ignored the neutrality of Bjerknes as a Norwegian citizen and involved his Leipzig institute and his assistants in the military weather service. In 1917 Bjerknes moved to Bergen as director of a newly founded institute of geophysics. In view of his experiences with pre-war and wartime meteorology in Germany he suggested to the Norwegian government to expand the Norwegian weather services. He would "gladly join in the processing of the observational data" (Friedman, 1989, pp. 103–108).

But not all meteorological work was kept secret. Franz Linke (1878–1944), a professor of meteorology at Frankfurt, had published in 1913 a book titled *The meteorological training of pilots* of which four years later a second enlarged edition appeared (Linke, 1917). It was praised by a reviewer because it presented "during the great surge of aeronautics in the war-years the many newly trained pilots with an introduction to the meteorological phenomena that are important for practical aviation".[10] The enlargements concerned in particular the "Turbulence of the air", as

[10] "... um während des großen Aufschwunges der Flugtechnik in den Kriegsjahren den zahlreichen neuausgebildeten Fliegern eine Einführung in die für die praktische Luftfahrt wichtigen meteorologischen Erscheinungen zu bieten." In: *Die Naturwissenschaften*, 27. April 1917, Heft 17.

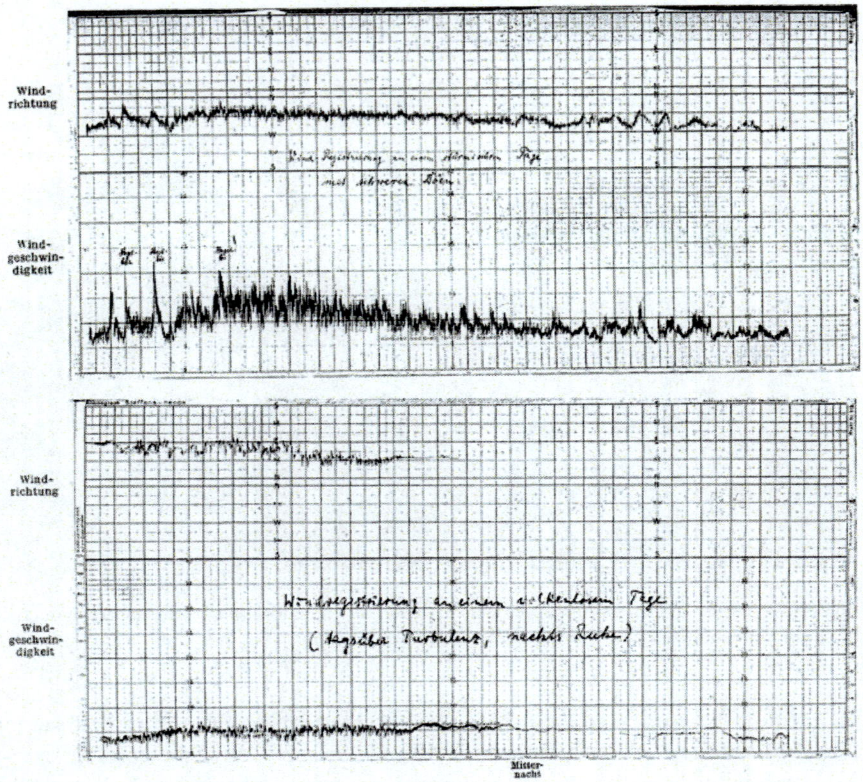

Fig. 4.3 Recordings of turbulent air in stormy (above) and quiet weather (Linke, 1917, Figs. 28 and 29)

a new section was titled in a chapter about "The motion of air and its disturbances". Pilots experienced turbulence like "a ride over a ploughed field" ("Fahrt über einen Sturzacker") (Linke, 1917, p. 54). But Linke did not content himself with a mere qualitative description. He added two readings of Pitot tube measurements which compared the air turbulence in stormy and quiet weather (Fig. 4.3).

Another non-secret report on turbulence concerned the wind fluctuations close to the ground measured at different heights of wireless transmission towers (Barkow, 1917). The author of this paper was Erich Barkow (1882–1923), a meteorologist who had acquired his expertise on turbulent winds as participant of the German Antarctic Expedition in 1911/12 (Barkow, 1924). In March 1917 he presented his views on turbulence also at a meeting of the Berlin branch of the Deutsche Meteorologische Gesellschaft. He coined the notion of a "turbulence body" for characterizing horizontal and vertical turbulent motions. The latter depended on the variation of temperature with increasing distance from the ground. "If both motions are equal, the turbulence body is a sphere; if the vertical temperature gradient decreases the turbulence body gets flattened." Together with wind measurements, therefore, attention should be paid to the distribution of vertical temperatures, for which he employed a thermoelectric method developed recently at the Meteorological Observatory in Potsdam. He wished that this method could be extended to the free atmosphere.[11]

With hindsight from the 21st century, Barkow's and other papers published in the early 20th century appear as early contributions to boundary layer meteorology—the modern term that covers most research on atmospheric turbulence [see e.g. (Kraus, 2008)]. At the time of the First World War, however, these terms were not yet in common use. The emergence of notions like "boundary layer" and other concepts for a theory of turbulence will be treated in the following part.

[11] "Sind beide Bewegungen gleich, so ist der Turbulenzkörper eine Kugel; wird der vertikale Temperaturgradient geringer, so plattet sich der Turbulenzkörper ab. [...] Eine Erweiterung dieser Methode auf die freie Atmosphäre wurde als wünschenswert bezeichnet." In: *Die Naturwissenschaften*, 30 March 1917, 5:13, p. 205.

Part II
Concepts

Turbulence research in the 20th century was preceded by engineering knowledge on pipe and channel flow, drag investigations, meteorological studies of atmospheric flow and other inquiries on eddying motion as described in Part I. As a rule, theoretical developments were guided by empirical knowledge. This part focuses on the formation of concepts for a theory of turbulence.

Chapter 5
The Boundary Layer Concept

Abstract In 1904 Ludwig Prandtl led the foundation for what became known as boundary layer concept. Already before the 20th century there were more or less developed views about a frictional layer between the solid surface of an object in a stream and the free flow at some distance from the surface, but none of these were further developed. Prandtl aimed with his concept at a theory of flow detachment from solid surfaces. In its initial form, the concept was elaborated for laminar boundary layer flow by Prandtl's doctoral student Heinrich Blasius. Turbulent boundary layer flow and detachment was observed at first in the Göttingen wind tunnel with experiments on the drag of spheres. In 1920, Prandtl's "master pupil" Theodore von Kármán paved the ground for a theory of turbulent boundary layers that became subject of a fierce rivalry between Prandtl, Kármán and their "schools" in Göttingen and Aachen. By the same time, the transition from laminar to turbulent flow, with the focus on the boundary layer, became perceived as "the turbulence problem". In 1924 hot-wire anemometry in a wind tunnel provided experimental evidence for the theoretically predicted laminar and turbulent flow profiles in the boundary layer. Both for the theory of the onset of turbulence and of fully developed turbulence the boundary layer concept proved crucial and fruitful.

5.1 Before Prandtl

Unlike other more gradually evolving concepts in science the boundary layer concept is associated with a sudden birth in 1904 and single originator, Ludwig Prandtl. Yet, at least qualitatively, it was presaged at numerous occasions long before. In his studies on pipe and channel flow, for example, Du Buat assumed that water would fill the gaps between the protrusions of the surface (Rouse and Ince, 1957, pp. 131–132). Therefore, rough and smooth walls could be treated on the same footing—a view that was opposed by another hydraulic engineer of the late 18th century, who argued that the water particles close to the wall would be "hindered by collisions, deviated into other directions and into an eddying motion whereby the entire flow is decelerated" (see Sect. 1.3), which reads like a description of turbulent boundary layer flow more than a hundred years before Prandtl. In 1865 Bazin's measurements

M. Eckert, *Turbulence—an Odyssey*, History of Physics,
https://doi.org/10.1007/978-3-030-91459-2_5

of velocity profiles of channel flow (see Sect. 2.3) provided clear evidence that wall roughness made a marked difference compared to smooth walls. In retrospect Bazin's velocity profiles may be regarded as approximations to those derived from turbulent boundary layer theory (Darrigol, 2005, p. 233).

Another hunch of boundary layers may be found in William Froude's study on ship resistance (see Sect. 3.4). When he determined in his towing tank the skin friction of submerged planks that varied in length from 10 inches to 50 feet, the drag did not increase in proportion with the length of the plank. Froude argued [quoted in Rouse and Ince (1957, pp. 185–186)]

> that the portion of surface that goes first in the line of motion, in experiencing resistance from the water, must in turn communicate to the water motion in the direction in which it is itself travelling; and consequently the portion of surface which succeeds the first will be rubbing, not against stationary water, but against water partially moving in its own direction, and cannot therefore experience as much resistance from it.

By the end of the 19th century the idea of a boundary layer was anticipated also from a more theoretical vantage point, such as in debates about the flow around a solid. "There will be an exceedingly thin stratum of fluid round the solid through which the velocity of the water varies continuously from the velocity of the solid to the velocity in the solution for inviscid fluid", Lord Kelvin argued in a letter to Gabriel Stokes in 1898 [quoted in Darrigol (2005, p. 205)]. The theoretical bone of contention was the question whether the results of ideal flow theory could be applied to viscous fluids in the limit of vanishing viscosity. In 1892 Lord Rayleigh (1842–1919) had speculated that his considerations on the instability of flows in inviscid fluids "may be found inapplicable to a viscid liquid of vanishing viscosity". Crucial were the processes "in the immediate neighbourhood of the walls, 'when the viscosity is very small'." [quoted in Darrigol (2005, p. 218)].

The concept of an atmospheric boundary layer, too, had early advocates in the 19th century. In a study on atmospheric motions published in 1876 the authors assumed a horizontal layer of air over the surface of the earth whose flow was determined by the balance of the Coriolis force and the friction on the ground. They compared the diminishing friction in the air layer with increasing height above ground with that observed in open channel flow: "As the height of the current increases, the value of k [i. e. friction coefficient] decreases, which is in accordance with what we know about the coefficient of water in open channels."[1] A few years later the author of a meteorological textbook further elaborated this view by taking the higher air layers into account which exerted a drag on the air underneath. He distinguished the "ground layer" ("Grundschicht") immediately adjacent to the surface from superposed middle and upper layers:

> All three layers influence each other by friction, whether it is the usual fluid friction known from the physics laboratory (i.e. the mutual exchange of molecules over very small distances)

[1] "Quand la hauteur du courant augmente, la valeur de k diminue, ce qui est conforme à ce que nous connaissons du coefficient de l'eau dans les canaux découverts." (Guldberg and Mohn, 1876, p. 19).

or the mass exchange caused, for example, by small ascending and descending air currents as a result of midday insolation.[2]

Hence there is no lack of more or less developed views that appear with hindsight like presage boundary layer concepts. None of these, however, was elaborated into a theory from which further applications could be derived. This step was left to Ludwig Prandtl and his budding Göttingen school.

5.2 The Genesis of Prandtl's Boundary Layer Concept

The boundary-layer concept was born in August 1904 when Prandtl presented to the Third International Mathematicians Congress in Heidelberg a paper "On fluid movement with very little friction" (Prandtl, 1905). It marked "the beginning of man's understanding of the dynamics of real fluids," as this event was praised retrospectively in a fifty-years-birthday article in *Science* (Dryden, 1955, p. 375). Another fifty years later, the centenary of this concept was celebrated with an international symposium. By then the scope of boundary layer research became evident by its accommodation in six sessions on the mathematics of boundary layers, instability of boundary layers and transition to turbulence, boundary layer control, turbulent boundary layers, numerical treatment and boundary layer modelling, and special effects in boundary layers (Meier and Sreenivasan, 2004). According to another centenary paper that emphasized the practical applications of the boundary layer concept it "has allowed prediction of skin friction drag, heat transfer from the wall and separation of the boundary layer, which in turn enables proper design of airplanes, ships, other vehicles and equipment through/around which fluid flow takes place." (Tulapurkara, 2005, p. 499).

In 1904, of course, such widespread ramifications and applications were still far in the future. Prandtl received the first incentive when he was employed after his study in 1900 as an engineer in a machine factory and trusted with the design of extraction conduits for removing shavings and dust from work benches (Eckert, 2017, Chap. 4). There he encountered the phenomenon of flow separation in diverging ducts which resulted in a loss of pressure. "In a large ventilating system in the machine factory in Nuremberg, I had arranged a conically enlarged pipe in order to regain pressure," Prandtl recalled many years later; "but, there was no regaining of pressure and a separation of the flow occurred."[3] When he was called in 1901 as professor of mechanics to the technical university in Hanover he further investigated the problem

[2] "Alle 3 Schichten beeinflussen einander durch Reibung, möge dieselbe nun in der gewöhnlichen, aus dem physikalischen Laboratorium bekannten Flüssigkeitsreibung (d. h. aas dem gegenseitigen Austausche der Moleküle auf sehr kleine Entfernungen) oder aber in dem Massenaustausche bestehen, wie derselbe z. B. infolge der mittäglichen Insolation durch kleine auf- und absteigende Luftströmungen verursacht wird." (Sprung, 1885, p. 126).

[3] "In einer größeren Luftleitungsanlage in der Maschinenfabrik Nürnberg hatte ich ein konisch erweitertes Rohr angeordnet, um dadurch Druck wiederzugewinnen; der Druckwiedergewinn ist aber ausgeblieben und dafür ist eine Ablösung der Strömung eingetreten." (Prandtl, 1948, p. 90).

Fig. 5.1 Ludwig Prandtl in front of the water trough from which he photographed the flow pictures for his boundary layer lecture at the Third International Mathematicians Congress in 1904 in Heidelberg. This picture became an icon of modern fluid mechanics. Credit DLR, GG-0010

of flow separation. He constructed a water trough with an intermediate floor which he provided with guide vanes at one end and a water wheel with a crank handle at the other, so that he could move the water in vertical circulation and observe the flow around objects arranged on the intermediate floor (Fig. 5.1). By adding iron mica to the water he visualized and photographed flow separation from the wall of various objects. These pictures became an important part of his Heidelberg presentation (Prandtl, 1905).

But Prandtl did not content himself with visualizing flow separation. As can be seen from his unpublished notes since summer 1903, he also approached flow separation from a theoretical vantage point. He conceptualized this process as an interplay between the free flow (assumed as frictionless) with vortex sheets that become detached from the wall of an object and pushed into the free flow, where the "decay of a vortex sheet in a frictionless fluid" would result in an eddying mixing of the entire flow.[4]

Thus he had in a first step to deal with the formation of a vortex sheet between the solid surface and the adjacent free flow. In a fluid with very small viscosity the velocity would rapidly increase from zero at the wall to its free flow value a short

[4] "Auflösung einer Wirbelfläche in einer reibungslosen Flüssigkeit. 9. Juni 1903"; see (Fig. 4.1).

distance away from the wall. All processes in which friction plays a role would take place within a narrow transition layer for which the Navier-Stokes equations could be simplified. For the simplest case, the flow along the surface of a smooth plate, Prandtl derived an approximate solution for the (laminar) friction of the plate. In his notes he marked this result with the date "29. II. 04".[5] This was the only case that he was able to solve approximately in terms of the simplified Navier-Stokes equations, he admitted at his Heidelberg presentation in August 1904, but he left no doubt that this was merely the beginning of a research effort on the way to a theory of flow detachment. For the time being he resorted to qualitative sketches (Fig. 5.2) and photographs from his water trough in order to illustrate where he was aiming to. "The most important result is the explanation of the formation of discontinuity surfaces (vortex sheets) at continuously curved boundaries", he summarized the essence of his Heidelberg presentation for the editors of the *Bulletin of the American Mathematical Society*. "Photographs of experiments are shown and interpreted with the help of theory."[6] In his Heidelberg presentation he expressed the same message in these words: "A fluid layer set in rotational motion by the friction at the wall moves into the free fluid and, exerting a complete change of motion, plays there a similar role as Helmholtz's' discontinuity sheets."[7]

In other words: Prandtl's Heidelberg presentation was rather the beginning of a research program on flow detachment than the communication of what was later called boundary layer theory. Turbulence was not mentioned explicitly as a topic for further investigation, but it was close when Prandtl regarded the boundary layer as an agency that effects a complete flow transformation in the wake of a curved body by detachment from its surface. The focus on the laminar boundary layer along a flat smooth plate is misleading because there is no flow detachment in this flow. Yet this was the only case that hints at the asymptotic approach without which Prandtl could not have obtained the friction formula and the velocity profile of the laminar boundary layer. When Prandtl was asked many years later why he had kept the mathematical aspects of his concept in his Heidelberg paper so short he answered that he had not been given enough time for presenting mathematical elaborations (Tani, 1977, pp. 91–92). Numerous pages of unpublished notes testify indeed to a considerable mathematical effort[8]—but it did not lead to a result that Prandtl could have used as an answer to his primary question about the flow detachment from a curved boundary. Therefore he resorted to qualitative arguments supported by sketches and photographs.

---

[5] SUB Göttingen, Cod. Ms. L. Prandtl, Nr. 14, no page number.

[6] "Als wichtigstes Ergebnis sei die Erkärung für die Entstehung von Trennungsflächen (Wirbelflächen) an stetig gekrümmten Grenzflächen erwähnt. Es werden Photogramme von Versuchen gezeigt und mit Hilfe der Theorie gedeutet." SUB Göttingen, Cod. Ms. L. Prandtl, Nr. 14, p. 43.

[7] "Es schiebt sich also eine Flüssigkeitsschicht, die durch die Reibung an der Wand in Rotation versetzt ist, in die freie Flüssigkeit hinaus und spielt dort, eine völlige Umgestaltung der Bewegung bewirkend, dieselbe Rolle wie die Helmholtzschen Trennungsschichten." (Prandtl, 1905, pp. 487–488).

[8] SUB Göttingen, Cod. Ms. L. Prandtl, Nr. 14.

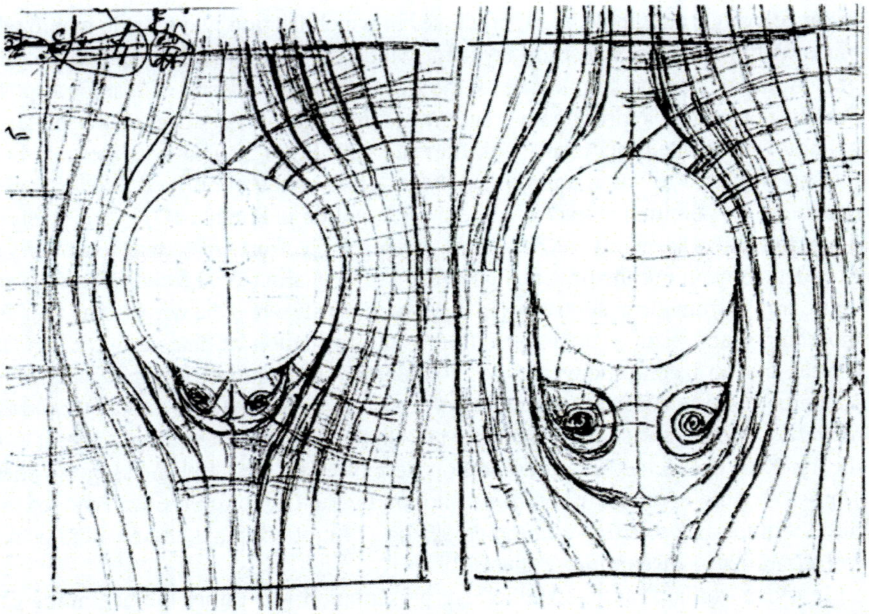

Fig. 5.2 Prandtl's sketch of discontinuity surfaces forming at the periphery of a cylinder (the flow is from above downwards) and rolling up in the wake. SUB Göttingen, Cod. Ms. L. Prandtl, Nr. 14, p. 38

5.3 Flow Detachment from Cylinders

In 1904 Prandtl was called to the university of Göttingen as professor of technical physics and director of an institute for applied mechanics. This offered him the opportunity to put his research program into practice. In 1910 he added heat transfer to the scope of phenomena where the flow in the immediate vicinity of solid boundaries plays a role. In this context he juxtaposed "two independent dimensionless quantities, namely the Reynolds number known in hydrodynamics, $\xi = \frac{\rho r \bar{w}}{k}$, and the ratio $\nu = \frac{\lambda}{ck}$", where r is the radius of a pipe through which a fluid of density ρ, viscosity k, thermal conductivity λ and specific heat c flows with the mean velocity $\bar{w}$ (Prandtl, 1910b). The ratio ν would later be called Prandtl number—to Prandtl's own annoyance because he credited the engineer Wilhelm Nusselt (1882–1957) with the discovery of its importance in heat transfer. In retrospect, Prandtl's paper is rather noteworthy because it assumed a laminar sublayer underneath a turbulent boundary layer, although this distinction was not made explicit (Rotta, 1981, p. 7).

While the boundary layer concept had no followers elsewhere before World War I, Prandtl made it a research subject of his budding school. His first doctoral student, Heinrich Blasius, solved the boundary layer equation for the laminar flow along a flat smooth plate ("Blasius flow") and computed the point at the periphery of a cylinder in a cross-flow where the (laminar) flow separates from the surface (Blasius, 1908; Hager, 2003). Subsequently Ernst Boltze considered these problems for bodies of revolution, such as airships, with their axis oriented parallel to the flow (Boltze, 1908). Another doctoral student, Karl Hiemenz (1885–1973), investigated the boundary layer of a cylinder in a cross-flow experimentally by measuring the pressure distribution along the periphery of the cylinder (Hiemenz, 1911).

Turbulence was hardly mentioned explicitly in these early boundary layer studies, yet it was not ignored as is evidenced by a seminar on hydrodynamics organized by the Göttingen mathematician Felix Klein in winter 1907/08 (Eckert, 2019a). The seminar was motivated by the challenge to bridge the gap between hydrodynamics and hydraulics. Next to Prandtl, Klein involved the applied mathematician Carl Runge (1856–1927) and the geophysicist Emil Wiechert (1861–1928) as co-organizers in this effort. Among the participants were four of Prandtl's doctoral students: Blasius presented a review on "turbulent flows", Hiemenz on "vortex motion", Georg Fuhrmann on "boundary layers and detachment of vortices", and Theodore von Kármán (1881–1963) on "unsteady potential motion". Blasius, for example, echoed Prandtl's view about the origin of turbulence when he discerned "the difference of flow velocities between neighbouring layers" as the cause of turbulence, "in the same way as Helmholtz's boundary layers become unstable". He also distinguished between "I: explanation of instability" and "II: Characterization of turbulent motion" as the two issues that hydrodynamics had to solve.[9]

In the aftermath of this seminar Klein paid himself tribute to the debate "On vortex formation in frictionless fluids". The debate had started with Helmholtz who used the creation of vortices in a cup of coffee by removing the spoon as an illustration for his vortex laws. Klein turned it into a thought experiment for the question what really causes the formation of vortices: friction or the change of topology before and after the removal of the spoon (Klein, 1910). "Klein's Kaffeelöffel experiment" became a generic case for the creation of vorticity (Saffman, 1992, Chap. 6).

For Hiemenz, who had anticipated some of Klein's arguments in his seminar presentation (Eckert, 2019a, pp. 89–91), vortex formation became a very practical problem in his doctoral work. When he attempted to measure the pressure distribution around a cylinder in a cross flow he observed small oscillations , "which are probably due to turbulence and other irregularities".[10] Turbulence apparently made a precise determination of vortex detachment difficult. "Prandtl struggled with the problem of solving the nature of these vortices," Kármán recalled many years later. "While I was

[9] "Die Entstehungsursache der Turbulenz haben wir in der Geschwindigkeitsdifferenz benachbarter Schichten zu suchen, etwa so, wie auch die Helmholtz'schen Grenzschichten instabil sind, weil einer Verlegung der Grenzschicht an der konkaven Seite eine Verringerung der Geschwindigkeit, also Erhöhung des Drucks entspricht, der auf Vergrößerung der Einbiegung hinwirkt. [...] Die Fragen, die durch diese soben erörterten Verhältnisse an die Hydrodynamik gestellt werden, sind: I Erklärung der Instabilität. II Darstellung der turbulenten Bewegung." Quoted in Eckert (2019a, pp. 130–131). Prandtl made the same distinction when he sketched a few years later a research program for turbulence (Eckert, 2017, Sect. 5.6).

[10] "Zu diesen Messungen ist zu bemerken, dass der Meniskus während der Beobachtungen nie ganz stille stand, sondern immer kleine Schwingungen ausführte, die wohl auf Turbulenz und andere Unregelmäßigkeiten zurückzuführen sind." (Hiemenz, 1911, p. 12).

at Göttingen, I was always aware that work on vortices was going on." The effort to determine the vortex separation from pressure measurements around the surface of a cylinder became the prelude to Kármán's own study about the stability of vortex arrangements in the wake of the cylinder (Karman, 1967, pp. 61–63):

> Unfortunately Hiemenz found that the pressures he measured always fluctuated. No matter how hard he tried, he couldn't remove this unsteadiness. The motion of the water was always oscillatory. [...] Every morning as I walked through the laboratory to my work I would pause and ask: "Herr Hiemenz, it oscillates still?" He replied sadly: "Ja. Always it oscillates." [...] I had a strange hunch that the solution was mathematical. [...] I assumed that the water rolled up into vortices as it passed over and under the cylinder and broke off into two trails, one from the top and one from the bottom. Then I considered two possibilities. First I assumed that the vortices are all symmetrical, that is, the top and bottom vortices form at the same time and start out together. [...] Then I made a second assumption in which the vortices are shed alternately from the top and bottom of the cylinder. [...] Or to put it another way instead of marching two by two, the vortices are staggered like lampposts along both sides of a street. [...] I went to Prandtl. He said very calmly: "So, you've found something. Very well, write it up, and I will present it."
>
> He did present it to the Göttingen Academy in 1911, and it was well received. In fact, it made my name internationally known in aeronautics circles.

What became known as "Kármán vortex street" had been investigated experimentally earlier by Henri Bénard (1874–1939) (Wesfreid, 2017) and others—Kármán's discovery concerned the analysis of flow stability and the computation of the drag associated with vortex shedding (Karman, 1911, 1912). The latter seemed paradoxical because Kármán's analysis was entirely based on ideal flow theory that did not allow the creation of vortices. Kármán resolved "this apparent contradiction" by reference to Prandtl's boundary layer theory which justified the use of ideal flow theory at some distance from the surface of the cylinder (Karman, 1911, p. 516). Kármán reviewed his early success in his autobiography as follows (Karman, 1967, p. 64):

> What is the importance of the vortex street? First it gives us a scientific picture of the structure of the wake behind a moving body under certain conditions. Second, when the structure of the wake is known, the form drag of a blunt body, such as a sphere or a cylinder, could be calculated accurately for the first time. With this information engineers had a clearer understanding of how to minimize drag by streamlining. This became a basic technique in the design of planes, ships, and, of course, racing cars.

It is therefore not accidental that Prandtl and his students regarded vortex shedding not only from the perspective of the boundary layer concept but also as a process that is pertinent for the investigation of drag. Fuhrmann, for example, paid great attention to form drag in his wind tunnel investigation of airship models: the difference between more or less streamlined models manifested itself by deviations of the pressure distribution in the wake from that expected by ideal flow theory (see Sect. 3.5). Like Kármán, Fuhrmann justified the use of ideal flow theory by reference to Prandtl's "detachment theory"—then a common label of the boundary layer concept:

> According to the detachment theory of Prof. Prandtl this is completely plausible, because at the rear end of the bodies, where the flow, retarded by friction, enters an area of higher pressure, the conditions for the detachment of the flow and the formation of vortices are given; these vortices correspond to what is called the wake of a ship.[11]

5.4 The Turbulent Boundary Layer

So far flow detachment was perceived in terms of Prandtl's laminar boundary layer flow concept from the year 1904. The extension towards a turbulent boundary layer theory began ten years later, motivated by the drop of the drag coefficient of spheres at a critical air speed that Eiffel had observed in his wind tunnel (Sect. 3.5). It led Prandtl to suspect that this effect was caused by the transition to turbulence in the boundary layer. He imagined that the eddies of the turbulent boundary layer entrain the quiet air at the surface of the sphere behind the point of detachment and the airstream attaches itself again to the sphere, "so that the point of detachment moves further rearwards until it finds a new equilibrium position; this results in a much smaller system of eddies and thus also in a smaller drag."[12] The trip wire experiment (Fig. 3.6) corroborated this view. Whatever triggered this transition—a trip wire, a rough surface of the sphere or the turbulence of the air stream in the wind tunnel—would result in a reduced wake and a drop of the drag coefficient.

It was obvious that this effect had important practical consequences. If such abrupt changes of the drag would occur with other bodies at speeds beyond those accessible in the wind tunnel, Prandtl feared a crisis of model testing. "Experiments on models, as conducted now everywhere, would be rendered worthless if the point of transition is located between the Reynolds numbers of the actual performance and those of the laboratory tests."[13] He therefore investigated a number of bodies of revolution in the wind tunnel at different Reynolds numbers, with and without grids that made the air stream turbulent. While there was no drop at all for a sharp disk, the effect occurred in a broad range of Reynolds numbers for more or less elongated ellipsoids (Fig. 5.3).

---

[11] "Nach der Ablösungstheorie von Prof. Prandtl ist dies auch vollkommen erklärlich, denn an dem hinteren Ende der Körper, wo die durch die Reibung verzögerte Strömung in ein Gebiet höheren Druckes eintritt, sind die Bedingungen für die Ablösung der Strömung und die Ausbildung von Wirbeln gegeben; diese Wirbel entsprechen dem, was man bei einem Schiff als Kielwasser bezeichnet." (Fuhrmann, 1912, p. 105).

[12] "... und der Luftstrom legt sich wieder an die Kugel an, so daß die Ablösestelle immer weiter nach hinten rückt, bis sie unter den neuen Bedingungen eine neue Gleichgewichtslage gefunden hat; dies hat ein wesentlich kleineres Wirbelsystem und damit auch einen kleineren Widerstand zur Folge." (Prandtl, 1914, p. 180).

[13] "Solche Umstände wären dazu angetan, die Versuche an Modellen, wie sie jetzt überall ausgeführt werden, wertlos zu machen, falls sich die Stelle des Umschlages zwischen den Reynoldsschen Zahlen der wirklichen Ausführung und denen der Laboratoriumsversuche befände." (Prandtl, 1914, pp. 186–187).

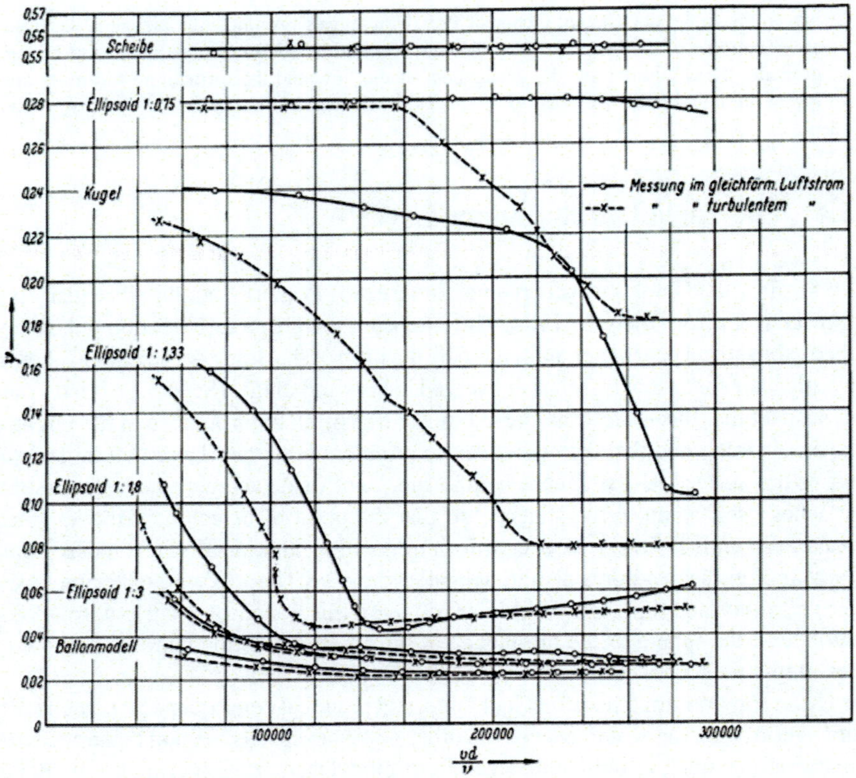

Fig. 5.3 The drop of the drag coefficient ψ for different rotational bodies (upper curve: disk) as a function of the Reynolds number. The dashed curves (x) were obtained with a wire grid that made the air stream turbulent; the circles (o) refer to the undisturbed air stream (Prandtl, 1914, Fig. 5)

Similar wind tunnel tests were made in World War I on streamlined struts with different profiles such as used for biplanes (Fig. 5.4). Prandtl's collaborator, Max Munk (1890–1986), summarized the results of these measurements in a wartime report. He recommended the use of strut profiles where the transition to turbulence in the boundary layer occurred already at very small speeds so that it would not happen during flight. "Only when this basic requirement is fulfilled one may think of optimizing the profile in order to further reduce the drag."[14]

Thus the turbulent boundary layer made its appearance in wartime applications long before it became a research subject of its own from a more fundamental perspective. Although Prandtl penned in 1916 on a sheet of paper a "Working program on the theory of turbulence" that addressed "Origin of turbulence" and "Developed turbu-

[14] "Erst wenn diese grundlegende Bedingung erfüllt ist, ist daran zu denken, durch besonders gute Linienführung des Profils den Widerstand noch weiter zu verkleinern." (Munk, 1917, p. 89).

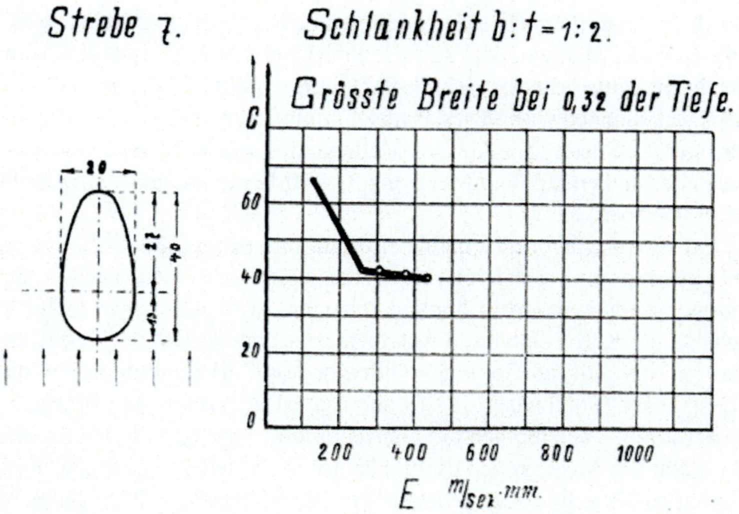

Fig. 5.4 The drop of the drag coefficient for a strut with a streamlined profile as used in biplanes in WW I (Munk, 1917, Tafel 43)

lence" as the two basic research goals,[15] there were more urgent items on his agenda during the war, such as a theory of aerodynamic lift, wind tunnel tests of wing profiles for airplane manufacturers and the expansion of the small airship model testing facility into a large aerodynamic testing establishment (Aerodynamische Versuchsanstalt) (Eckert, 2017, Sect. 4.1).

The execution of Prandtl's "working program" on turbulence had to wait for the time after the war. As soon as professors and students returned from the trenches to the universities, turbulence became part of the research agenda—not only in Prandtl's institute. "I was informed by [Karl] Pohlhausen [1892–1980] that you will lecture on turbulence at Nauheim," Prandtl wrote to Kármán in August 1920, a few weeks prior to the first postwar meeting of the Association of German Natural Scientists and Medical Doctors (Gesellschaft deutscher Naturforscher und Ärzte) in Bad Nauheim.[16] The annual meetings of this association provided an opportunity for applied mathematicians, physicists, and engineers in German speaking countries to address common concerns. The meeting at Bad Nauheim marked the transition into a new era, from a scientific vantage point as much as from the perspective of science organization in Weimar Germany (Forman, 2007). In its aftermath Prandtl, Richard von Mises, Kármán and other application oriented scientists voiced the need for a new umbrella for their common concerns. Industrial physicists founded a new society, the German Society for Technical Physics (Deutsche Gesellschaft für Technische Physik), in close association with the older German Physical Society (Deutsche Physikalische

[15] Dated 6 March 1916. SUB, Cod. Ms. L. Prandtl, Nr. 18.

[16] "Durch Pohlhausen hörte ich, dass Sie in Nauheim über Turbulenz vortragen werden." Prandtl to Kármán, 11 August 1920. GOAR 1364.

Gesellschaft). In September 1921, these two physical societies held a joint meeting with the German Mathematical Association (Deutsche Mathematiker-Vereinigung) in Jena. In the same year, the *Zeitschrift für Angewandte Mathematik und Mechanik (ZAMM)* was founded with Mises as its longtime editor (Siegmund-Schultze, 2020). A year later the Society for Applied Mathematics and Mechanics (Gesellschaft für Angewandte Mathematik und Mechanik, GAMM) was established, with Prandtl as president (Gericke, 1972).

Prandtl's inquiry about Kármán's effort on turbulence, therefore, was motivated not only by scientific curiosity: it reflects the beginning of their rivalry in the context of a growing appreciation of applied sciences—with turbulence as the predominant challenge. Karl Pohlhausen was their common student resp. collaborator. He had finished his study in Göttingen with a doctoral dissertation on laminar boundary layer theory (Pohlhausen, 1920) and moved to Aachen as Kármán's assistant in January 1920. The "momentum consideration" employed by Pohlhausen for the laminar boundary layer proved useful also for the turbulent boundary layer, as Kármán remarked in a five-page-letter to Prandtl in February 1921 about his recent effort to conceive "a kind of 'turbulent boundary layer theory'". Conscious about the competitive nature of his attempt he added the source of his inspiration, an earlier conversation with Prandtl on skin friction, and the beginning of his own elaboration, "December 1920".[17] Prandtl swiftly acknowledged Kármán's priority. "In the general formulation of the 'turbulent boundary layer theory' you are decidedly further advanced than I am, because I had only planned something like this for the more distant future and had not yet begun to elaborate it."[18]

The goal of Kármán's effort was the "wall law". It describes the growth of mean velocity u in the turbulent boundary layer as a function of the distance y from the wall. Prandtl had derived this law in the form $u \sim y^{1/7}$, as Kármán recalled, but he could not reconstruct Prandtl's argument from his memory and therefore asked him "whether you have already published your 1/7 law" so that he could refer to it. Furthermore he was curious whether his own derivation was identical to that of Prandtl.[19] Another doctoral student of Prandtl, Johann Nikuradse (1894–1979), later recalled that Prandtl had derived this law on 5 November 1920 in a seminar; Prandtl's "consideration" started with the empirical law of Blasius (see Sect. 1.7) combined with the assumption

[17] "Sie haben mir in der Rhön von Ihrem $y^{1/7}$-Gesetz erzählt [...] ich habe mir die Sache danach rekonstruiert und auf dieser Grundlage eine Art 'turbulente Grenzschichttheorie' aufgebaut [...] was ich mir über die Plattenreibung zurechtgelegt habe (Dezember 1920) [...] Nun habe ich dieselbe Methode der Impulsbetrachtung angewendet, welche ich für Herrn Pohlhausen vorgeschlagen habe und welche sich für laminare Grenzschicht sehr gut bewährt hat." Kármán to Prandtl, 12 February 1921. GOAR 3684.

[18] "In der allgemeinen Formulierung der 'turbulenten Grenzschichttheorie' sind Sie entschieden weiter als ich, denn ich hatte mir so etwas nur für die fernere Zukunft vorgenommen gehabt und war an die Ausarbeitung noch nicht herangegangen." Prandtl to Kármán, 16 February 1921. MPGA, Abt. III, Rep. 61, Nr. 792.

[19] "Ich möchte Sie jedoch bitten, mir mitzuteilen, ob Sie Ihr 1/7 Gesetz bereits veröffentlicht haben, so dass ich mich darauf beziehen kann, oder dass Sie es demnächst veröffentlichen wollen. Ferner ist meine Ableitung mit Ihrer Ableitung identisch oder nicht?" Kármán to Prandtl, 12 February 1921. GOAR 3684.

that the shear stress at the wall was only dependent on the velocities close to the wall, but not on the pipe diameter, and that the velocity distribution as a function of the distance from the wall y and the radius r followed a power law $u(y) = u_0(\frac{y}{r})^q$, where u_0 is the velocity at the centerline of the pipe and q an unknown exponent. By dimensional analysis—the derivation required little more than a single page—he obtained the result $q = 1/7$ (Nikuradse, 1926, pp. 14–15). There was no further assumption on turbulence that would justify to call this consideration a theory of boundary layer turbulence. Prandtl therefore frankly admitted Kármán's priority with regard to a "turbulent boundary layer theory" and encouraged him to proceed with his plans. "Ultimately, I can get over it if the precedence of publication has gone over to friendly territory."[20]

The wider community of applied scientists learned about these efforts from Kármán's and Pohlhausen's papers that appeared in subsequent articles in August 1921 in *ZAMM* (Karman, 1921; Pohlhausen, 1921). The major conceptual tool of Kármán's theory was the "momentum consideration" which exposed the "physical meaning" of the boundary layer concept and made it possible to account for both laminar and turbulent flow in the boundary layer in terms of relatively simple calculations. "As we shall see, we arrive at laws of turbulent friction on a dragged plate, which correspond very well with experience", Kármán introduced his derivation of the 1/7-law as a major success of his theory. In addition to the velocity profile in the turbulent boundary layer he derived consequences for the heat transfer which "considerably extended" Prandtl's "analogy conclusions" from the year 1910. He also derived by dimensional analysis the friction factor for rough walls, using the concept introduced by Mises in 1914 (Sect. 1.7). "Mister Prandtl tells me that he has obtained the same result in an entirely different way," he added in a footnote.[21]

5.5 The Turbulence Problem—as Perceived in the Early 1920s

Kármán's and Prandtl's (yet unpublished) approach on boundary layer turbulence was based on the empirical law of Blasius for pipe flow. From the perspective of Prandtl's "working program" it fell under the category of "developed turbulence". The other part of the program, "origin of turbulence", also became subject of great efforts. A review published in *ZAMM* under the headline "The turbulence problem" character-

[20] "Ich werde dann hinterher schon sehen, wie ich mit meiner anderen Herleitung noch zu meinem Recht komme und werde es schliesslich verschmerzen können, wenn die publizistische Priorität in befreundeten Besitz übergegangen ist." Prandtl to Kármán, 16 February 1921. MPGA, Abt. III, Rep. 61, Nr. 792.

[21] "Wir gelangen so, wie sich zeigen wird, zu Gesetzmäßigkeiten der turbulenten Reibung an einer geschleppten Platte, die mit der Erfahrung sehr gut übereinstimmen. [...] erscheint die Berechnungsmöglichkeit der Wärmeübertragungsvorgänge über die erwähnten Prandtlschen Analogieschlüsse hinaus wesentlich erweitert [...] Herr Prandtl teilt mir mit, daß er dasselbe Ergebnis auf ganz anderem Wege erhalten hat." (Karman, 1921, pp. 237, 242, 251).

ized earlier attempts to account for the onset of turbulence in terms of stability theory as a complete failure (Noether, 1921). This approach had been elaborated in 1907 by the Irish mathematician William McFadden Orr (1866–1934) and in 1908, independently, by Arnold Sommerfeld. The reviewer, Fritz Noether (1884–1941), was a student of Sommerfeld and had paid tribute to this problem with detailed studies. The goal of the "Orr-Sommerfeld approach", as it became known, was to derive from an eigenvalue equation the critical Reynolds number beyond which the laminar flow becomes unstable when a wavelike disturbance was superposed to it. As main flow the simplest case was assumed, a plane flow between two walls moving in opposite directions (plane Couette flow). The result was always the same: The flow remained stable at all Reynolds numbers and all wavelengths of the superposed disturbance. There was no critical limit beyond which the flow would become turbulent (Eckert, 2010).

Noether's review in *ZAMM* made the "turbulence problem" an issue of controversial debates among physicists, mathematicians and "scientific engineers" like Prandtl, Kármán and others of that ilk. At the joint meeting of the German Society for Technical Physics, the German Physical Society and the German Mathematical Association, held in September 1921 in Jena, Prandtl presented the preliminary results that his doctoral student, Oskar Tietjens (1893–1971), had obtained on the stability of laminar flow. In contrast to the Orr-Sommerfeld approach which took into account the whole range of Reynolds numbers for the stability analysis, Tietjens started from the inviscid limit (i. e. Reynolds numbers $\longrightarrow \infty$) because the instability that would trigger the onset of turbulence was to be expected only at high Reynolds numbers. Therefore he based his analysis on an equation derived by Lord Rayleigh for the stability of plane parallel inviscid flows (Darrigol, 2005, Sect. 5.5). "We, that is in particular Mr. O. Tietjens who conducted the calculations under my supervision, investigated the stability and instability of laminar flows as they form along a wall under the influence of a small viscosity, according to Lord Rayleigh's method in the limit of vanishing friction," Prandtl introduced the work of his doctoral student at the Jena conference.[22] As expected from Rayleigh's analysis, any inflection of the profile (like c in Fig. 5.5) would lead to instability in the limit of infinitely high Reynolds numbers.

The surprise came when Prandtl and Tietjens extended the analysis by introducing a small viscosity. Then even profiles without an inflection became unstable! "At first we did not want to believe this result, so we each calculated it three times independently in different ways. The same sign came up again and again, which means lability." Whereas the Orr-Sommerfeld approach ended in the dead-lock of stability, now they faced "the opposite difficulty, how to save the laminar flow".[23]

---

[22] "Wir, d. h. wesentlich Herr O. Tietjens, der unter meiner Leitung die Rechnungen machte, untersuchten die Stabilität und Labilität von Laminarströmungen, wie sie längs einer Wand durch länger dauernde Einwirkung einer geringen Zähigkeit entstehen, und zwar nach der von Lord Rayleigh angegebenen Methode unter Vernachlässigung der Reibung." (Prandtl, 1922, p. 19).

[23] "Wir haben dieses Ergebnis erst nicht glauben wollen und haben es daraufhin jeder dreimal unabhängig auf verschiedene Weise gerechnet. Es kam immer wieder dasselbe Vorzeichen, das Labilität

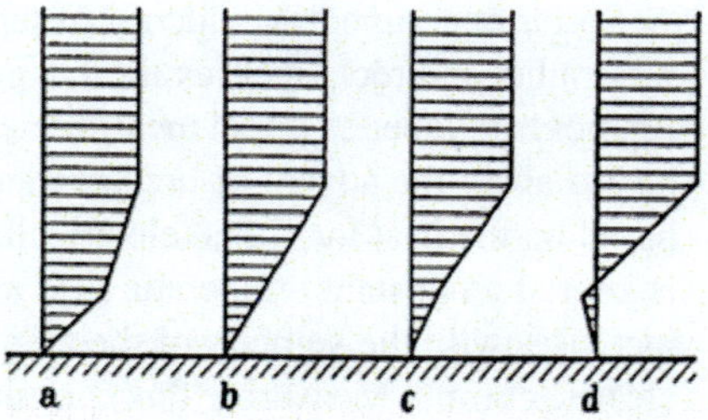

Fig. 5.5 The profiles analysed by Prandtl and Tietjens for stability according to Rayleigh's approach (Prandtl, 1922, Fig. 1)

The discussion about these results reflected a state of confusion. Sommerfeld doubted Prandtl's view that plane Couette flow was different from all other cases with regard to stability. Kármán blamed the straight-line profiles with kinks at certain points for the flawed stability analysis. He also expressed doubts about Prandtl's conviction that closer experimental observations would help to understand the nature of turbulence. "What one observes are mostly the immensely complicated processes in the transition region between laminar and turbulent flow, influenced by all kinds of accidence. The stable turbulent flow in statistical equilibrium eludes direct observation because the processes proceed much too fast."[24]

Nevertheless, in view of the paradoxical results of theoretical studies about the onset of turbulence, both Kármán and Prandtl resorted to experimental investigations. Ludwig Schiller (1882–1961), an experimental physicist at the university of Leipzig, spent research sojourns in Prandtl's institute in order to explore the role of disturbances at the entrance of a pipe for the transition to turbulence. By an appropriately shaped mouth and an inflow that avoided disturbing motions as far as possible, he was able to maintain laminar flow up to a critical Reynolds number 9600. On the other side, an eddying inflow caused turbulence at a lowest critical Reynolds number of 1160. "For each Reynolds number above 1160, there is a specific amount of disturbance required to cause turbulence," Schiller concluded. "The higher the Reynolds number, the smaller the disturbance that is sufficient."[25] Experimental investigations on the influence of wall roughness on turbulent pipe flow were conducted in Kármán's laboratory at the technical university in Aachen. With regard to the "turbulence problem", i.e. the transition from laminar to turbulent flow, the Aachen experiments specified the limiting Reynolds numbers for the transition region between 500 and 8000, but excluded entrance disturbances as the cause for this wide range (Fromm, 1923, p. 20).

---

bedeutet. [...] ... denn es entstand jetzt die umgekehrte Schwierigkeit, wie die Laminarströmung gerettet werden sollte ..." (Prandtl, 1922, pp. 21–22).

[24] "Was man beobachtet, sind zumeist die ungemein komplizierten, durch allerlei Zufall beeinflussten Vorgänge im àbergangsgebiet zwischen laminarer und turbulenter Strömungsform. Die im statistischen Gleichgewicht befindliche stabile turbulente Strömungsform entzieht sich zunächst der direkten Beobachtung, weil die Vorgänge viel zu rasch vor sich gehen." Kármán in (Prandtl, 1922, pp. 24).

[25] "Zu jeder Reynolds Zahl oberhalb 1160 gehört ein ganz bestimmter Störungsbetrag, der erforderlich ist, um die Turbulenz hervorzurufen. Je höher die Reynoldssche Zahl ist, eine um so geringere Störung reicht hierzu aus." (Schiller, 1921, p. 443).

So far experimental evidence of laminar or turbulent flow in the boundary layer was rather indirect, such as the change of the friction coefficient with increasing Reynolds number. A direct measurement of the velocity profile in the boundary layer had to await the advent of hot-wire anemometry. The principle of this technique is based on the heat loss of an electrically heated wire in an air stream. If the hot-wire is kept at a constant temperature (at a given electrical resistance) the heat loss that increases with the velocity of the air stream must be compensated by an increasing electric current. Measuring this current thus allowed to determine the velocity of the air stream.

The first use of this technique for boundary layer measurements was made in the mid1920 s in a wind tunnel of the laboratory of Johannes Martinus Burgers (1895–1981) at the technical university in Delft. The measurements were conducted on a flat smooth plate, using thin wires of 0,005 cm diameter and air speeds from 400 to 2400 cm/s. The results confirmed "the approximate agreement with Blasius's formula in the laminar part of the boundary layer, and the satisfactory verification of von Kármán's relation $u \sim y^{1/7}$ in the turbulent part". Thus Burgers summarized in 1925 the first direct confirmation of the boundary layer concept. The transition region from laminar to turbulent boundary layer flow was more difficult to determine because of large velocity fluctuations; these were "so intense that they could be observed by the eye from the glittering of the wire." Nevertheless it was possible to record "a very rapid increase of the thickness of the boundary layer", with an approximate relation of 1:1,4 for the thickness of the laminar to the turbulent boundary layer. "It may be expected that a still more detailed study of this region, and especially of the fluctuations of the velocity, may give valuable information about the nature of the vortices which mark the first appearance of turbulent motion." (Burgers, 1925, pp. 124–128).

Chapter 6
The Mixing Length

Abstract The mixing length concept emerged from efforts to solve the equations for mean turbulent motions established by Boussinesq (1877) and Reynolds (1895). Boussinesq's eddy viscosity was related to the shear stress exerted by fluctuating velocities (Reynolds stress) by means of a characteristic length which was interpreted by Prandtl as a "braking distance" along which turbulent bales of fluid transfer momentum to the ambient fluid. Kármán regarded this length as a characteristic scale of turbulent velocity fluctuations. For Taylor it represented a measure of turbulent mixing by an exchange of vorticity. The rivalry between Prandtl and Kármán about a "wall law" was at the same time a competition about the mixing length concept for which Kármán claimed to have developed the more universal formulation (involving a universal constant, called now the Kármán constant).

6.1 The Eddy Viscosity—A Key to Turbulence

The computation of turbulent flows from first principles had to start from the Navier–Stokes equations. Reynolds published in 1895 a paper in which he decomposed the flow into mean and fluctuating parts. Thus he obtained from the Navier–Stokes equations averaged momentum equations that involved terms with products of fluctuating velocities (Reynolds stresses) (Reynolds, 1895). However, this approach suffered from the closure problem: it involved more unknowns (the velocity fluctuations) than available equations, so that additional equations were required for the calculation of turbulent flows.

Two decades earlier, Boussinesq had aimed at the computation of turbulent pipe- and channel flow (see Sects. 1.6 and 2.4) with a similar approach. Unlike Reynolds, however, Boussinesq assumed that in fully turbulent flow the molecular viscosity ν should be replaced by a variable friction coefficient ε due to the turbulent motion (Boussinesq, 1877). Thus was born the concept of an eddy viscosity—although Boussinesq did not use that term. Boussinesq's approach also met with the closure problem because of the unknown eddy viscosity, ε; but this quantity appeared more accessible than Reynolds velocity fluctuations because it lent itself for physical assumptions in specific flow configurations.

M. Eckert, *Turbulence—an Odyssey*, History of Physics,
https://doi.org/10.1007/978-3-030-91459-2_6

With hindsight, Boussinesq's friction coefficient ε already contained the seed for the mixing length approach. By its dimension it was analogous to the molecular viscosity ν, but unlike the latter it was not a constant. Boussinesq suggested that for channel flow ε is roughly proportional to the mean flow velocity and a characteristic length,[1] which represents "already a mixing length formulation, as developed later by Prandtl" (Schmitt, 2007, p. 620).

Whereas the Reynolds equations were largely ignored as a viable gateway to the computation of turbulent flows, Boussinesq's approach was regarded as a promising concept. In 1904 the participants of Klein's seminar on hydrodynamics attempted to determine its potential for a theory of turbulent pipe- and channel flow (Hahn et al., 1904). They regarded ε so crucial that they called it "turbulence" (see Sect. 2.4). Their approach was further discussed in a subsequent seminar three years later by Blasius (Eckert, 2019a, pp. 134–135)—but none of these discussions were elaborated into a mixing length concept.

In 1915 Taylor explicitly introduced the "eddy viscosity" (without reference to Boussinesq) as a concept in order to account for turbulent atmospheric flows (see Sect. 4.3). He came close to a mixing length approach when he discerned the following relation for further study:

$$\frac{\mu}{\rho} = \frac{1}{2}(\bar{w}d).$$

Here μ is the eddy viscosity, d the average distance through which an eddy moves before it mixes with the surrounding fluid (density ρ) and $\bar{w}$ is the average eddy velocity. d may be regarded as a mixing length, although Taylor did not use this term. Nevertheless he perceived the eddy viscosity as the key for a theory of turbulent mixing because it raised the prospect to relate it directly to empirical data. He hoped that "if any method of deducing μ from meteorological observations could be found, it would be possible to verify the relation numerically" (Taylor, 1915, p. 14).

Next to Boussinesq and Taylor, Schmidt opened with his "exchange quantity A" (see Sect. 4.3) another route for the computation of turbulent mixing. It provided the link between hydraulics and meteorology for Prandtl and his doctoral student, Walter Tollmien (1900–1968), in a paper on "The distribution of wind over the ground, computed from the laws of pipe flow" (Prandtl and Tollmien, 1925). They introduced their subject as follows: "The recently very refined knowledge of the flow through pipes with rough walls gives the possibility, with the help of certain analogies, to state details about the wind motion over the rough earth ground".[2] Preceding measurements from Kármán's laboratory of pipe flow at different degrees of wall roughness (Hopf, 1923) had suggested to Prandtl and Tollmien an application to the flow of air over different terrain. They accounted for the turbulent mixing in terms of Schmidt's A by setting it equal to $\rho\varepsilon$, where ε is the "kinematic exchange"

---

[1] "... ε n'étant qu' à peu près proportional à u_0 et au rayon moyen h" (Boussinesq, 1877, p. 50).

[2] "Die in letzter Zeit sehr verfeinerten Kenntnisse von der Strömung durch Rohre mit rauher Wand geben die Möglichkeit, mit Hilfe gewisser Analogien Einzelheiten über die Windbewegung über dem rauben Erdboden auszusagen." (Prandtl and Tollmien, 1925, p. 47).

in analogy to the "kinematic viscosity".[3] The shear stress in the turbulent wind flow thus became

$$\tau = A\frac{\partial u}{\partial y} = \rho\varepsilon\frac{\partial u}{\partial y},$$

where u is the horizontal wind velocity and y the height above ground. With ε and u chosen in analogy to pipe flow with different wall roughness they arrived at formulae for the distribution of wind speeds over different terrain.

Although there was no mixing length involved in this approach, it offered the prospect for elaborating other cases of turbulent mixing.

6.2 Prandtl's Mixing Length Approach

In the averaged Navier–Stokes equations, as formulated by Reynolds in 1895, turbulent shear stress is represented by the Reynolds stress tensor

$$\tau_{ij} = \rho\langle u_i' u_j' \rangle,$$

where u_i' and u_j' are velocity fluctuations in the direction of x_i and x_j (with $i, j = 1, 2, 3$). For a two-dimensional flow ($u_1 = u = \bar{u} + u'$; $u_2 = v = \bar{v} + v'$, $u_3 = w = 0$) we have

$$\tau_{xy} = \rho\langle u'v' \rangle.$$

Comparing this representation with the shear stress derived from Schmidt's "exchange quantity" suggests that $\varepsilon\frac{\partial\bar{u}}{\partial y}$ is the product of two velocities. Because ε may be represented as a product of a length with a velocity, Prandtl made the Ansatz $u' = v' = l\frac{\partial\bar{u}}{\partial y}$, from which he obtained

$$\tau_{xy} = \rho l^2 \left(\frac{\partial\bar{u}}{\partial y}\right)^2.$$

He called the length l a "mixing path" ("Mischungsweg") in analogy to the mean free path of the kinetic gas theory. Prandtl regarded it as a characteristic measure for the turbulent flow under consideration. "The determination of l from the experimental results provides a particularly simple indication of the state of turbulence which, in conjunction with the equation for τ, provides everything essential about the apparent frictional forces inside the turbulent fluid." In this way Prandtl introduced the mixing

[3] "… und die wir mit Schmidt 'Austauschgröße' nennen wollen. Wir setzen diese = $\rho\varepsilon$, wobei dann ε der 'kinematische Austausch' wird, der genau analog der 'kinematischen Zähigkeit' ν ist …" (Prandtl and Tollmien, 1925, p. 49).

length approach in June 1925 at a meeting on hydraulics in Göttingen.[4] A few weeks before he had delivered a similar presentation at a meeting of engineers in Dresden. At this occasion he likened l to the "braking distance" ("Bremsweg") of a "bale of fluid" ("Flüssigkeitsballen") before it becomes mixed in the ambient fluid. He also mentioned a first successful application of this approach to a case of "free turbulence", the broadening of a jet of air that mixes with the ambient quiet air. "For stationary flows of this type, the approach $l = cx$ proves useful, where x is the distance from the point where the mixing begins."[5]

Prandtl's close collaborator for these investigations was Tollmien. They started with the elaboration of the mixing length concept shortly after their common paper for the *Zeitschrift für Geophysik* in autumn 1924 (Prandtl and Tollmien, 1925) in which they had invoked Schmidt's exchange concept. At that time Prandtl had informed Kármán about his further plans:

> I myself have occupied myself recently with the task to set up a differential equation for the mean motion in turbulent flow, which is derived from rather simple assumptions and seems appropriate for very different cases. We are now busy with its numerical elaboration and the computation of channel flow cases as well as such of the broadening of free jets which then should be compared with experiments. The beginning is quite promising. The empirical is condensed in a length which is entirely adjusted to the boundary conditions and which plays the role of a free path length.[6]

In his first report on the mixing length approach Prandtl announced a separate paper by Tollmien about the computation of these cases; he contented himself with the statement that the agreement with experimental results was "excellent" ("vorzüglich") (Prandtl, 1925, p. 138). The details were published in the following year by Tollmien. For a plane jet in x-direction (in first approximation at constant pressure) that broadened laterally in y direction the differential equation for the motion in x-direction for the mean velocities is

$$\bar{u}\frac{\partial \bar{u}}{\partial x} + \bar{v}\frac{\partial \bar{u}}{\partial y} = \frac{1}{\rho}\frac{\partial \tau_{xy}}{\partial y}.$$

Inserting the above expression for τ_{xy} with $l = cx$, where x is the distance from the opening and c a constant, yields a partial differential equation that could be solved

---

[4] "Die Ermittlung von l aus den Versuchsergebnissen liefert eine besonders einfache Angabe über den Turbulenzzustand, die in Verbindung mit der Gleichung für τ alles Wesentliche über die scheinbaren Reibungskräfte im Inneren der turbulenten Flüssigkeit liefert." (Prandtl, 1926a, p. 10).

[5] "Bei stationären Strömungen dieser Art bewährt sich der Ansatz $l = cx$, wo x die Entfernung von der Stelle ist, wo die Vermischung beginnt." (Prandtl, 1925, p. 138).

[6] "Ich selbst habe mich in der letzten Zeit viel mit der Aufgabe beschäftigt, für die mittlere Bewegung einer turbulenten Strömung eine Differentialgleichung aufzustellen, die aus ziemlich plausiblen Annahmen abgeleitet wird, und für sehr verschiedene Fälle geeignet erscheint. Wir sind gerade daran, sie numerisch zu behandeln und sowohl Fälle der Kanalströmung, wie solche der freien Strahlausbreitung zu rechnen, die dann mit dem Experiment verglichen werden sollen. Der Anfang ist ganz verheissungsvoll. Das Empirische wird dabei ganz auf eine den Randbedingungen angepasste Länge, die der freien Weglänge entspricht, gebracht." Prandtl to Kármán, 10 October 1924. MPGA, Abt. III, Rep. 61, Nr. 792.

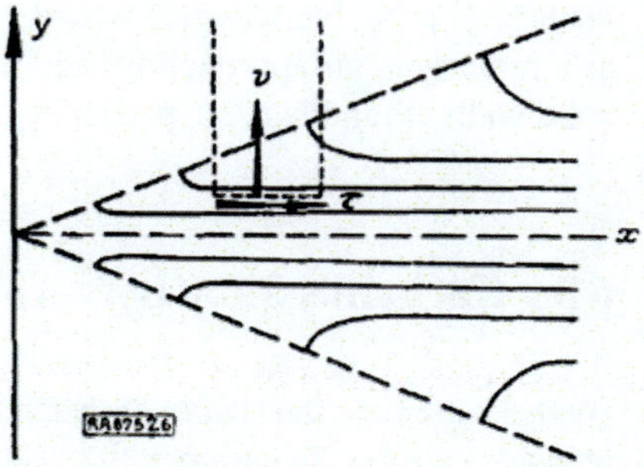

Fig. 6.1 The streamlines in two-dimensional jet broadening (the jet is emitted along the x-axis through a narrow slit at $x = 0$) computed by the mixing-length approach (Tollmien, 1926, p. 471)

by similar methods (stream function, substitution of variables) like the boundary layer equation. Besides the mixing of a homogeneous stream of air with the ambient quiet air, as it occurs in wind tunnels with an open test section, Tollmien was able to compute the spreading of a plane jet emitted through a narrow slit (Fig. 6.1) and a round jet through a hole (Tollmien, 1926).

In summer 1926 Prandtl presented the mixing length concept at the Second International Congress for Technical Mechanics in Zurich as a first step in his struggle with "the big problem of developed turbulence". He characterized this effort as "phenomenological" because the mixing length had to be chosen in agreement with plausible assumptions about the specific cases. The cases treated so far were plane or rotational-symmetric flows. Three-dimensional problems proved elusive for the mixing length approach. This was part of "the big turbulence problem" for which Prandtl had no answer:

> In any case, this phenomenon clearly shows us that the developed turbulence is an essentially three-dimensional motion. This circumstance, however, seems to move the solution of the big turbulence problem very far away, because unfortunately our present mathematical tools are generally quite inadequate for three-dimensional fluid movements.[7]

Yet the problems solved by Tollmien were not the only cases amenable to the mixing length approach. Another doctoral student of Prandtl, Hermann Schlichting (1907–1982), solved the problem of the "plane wake flow" ("ebene Nachlaufströmung") that occurs for example in the wind shadow behind a cylindrical body (Schlichting, 1930).

Prandtl also attempted to reconstruct the velocity distribution in the turbulent boundary layer in terms of the mixing length concept. But the analogous approach used for the free jet did not yield the expected $U \sim y^{1/7}$ distribution of the mean velocity U at the distance y from the wall. In a public lecture in Tokyo in October 1929 Prandtl concluded that "l proportional y does not lead to the desired result because it

[7] "Jedenfalls zeigt uns diese Erscheinung deutlich, daß die ausgebildete Turbulenz eine wesentlich dreidimensionale Bewegung ist. Dieser Umstand scheint allerdings die Lösung des großen Turbulenzproblems in sehr weite Ferne zu rücken, denn den dreidimensionalen Flüssigkeitsbewegungen gegenüber sind unsere heutigen mathematischen Hilfsmittel leider im allgemeinen recht unzureichend." (Prandtl, 1927, p. 75).

leads to U prop. log y, which would yield $U = -\infty$ for $y = 0$." He therefore aimed at a more general approach for the mixing length when the flow was bounded by solid walls (Prandtl, 1930, p. 9).

6.3 Kármán's Similarity Approach

By the same time the rivalry between Prandtl and Kármán for a universal "wall law" reached a climax. In summer 1928 Kármán informed Prandtl that he planned similar experiments on channel flow in his Aachen institute like Nikuradse in Göttingen:

> With regard to hydraulic questions we naturally also want to extend the experiments with smooth and rough channels to large Reynolds numbers. I have installed two large vessels for this purpose and a water tower. However, I believe that a duplication of work is not harmful for these basic experiments, especially because one wants to use the experiments as a guide for working out the theory.[8]

Walter Fritsch (1899–1987), whom Kármán trusted with these experiments as his doctoral student, found that the velocity profiles of different flows line up with each other around the centerline of the channel if they are shifted parallel. The same profile could result from the flow along a smooth wall at small speed and one at larger speed along a wall with a rough surface (Fritsch, 1928). Kármán inferred from this observation that the velocity distribution depends only on the shear stress transferred to the wall—disregarding the particular wall surface structure. In a letter to Burgers he revealed his further steps towards a new theory of turbulent skin friction:

> I have also elaborated what may be said about the resistance law if one makes the following assumptions:
>
> (a) The velocity profile in similar pipes only depends on the shear stress at the wall.
> (b) The mixing length in Prandtl's sense close to the wall is proportional to the distance from the wall.
> (c) For the transition to the laminar sublayer one replaces in Prandtl's approach τ by the quantity $\tau - \mu \frac{dU}{dn}$.
>
> From these assumptions one arrives at the resistance law [...] The only essential constant hereby is the proportionality factor of the mixing length close to the wall.[9]

---

[8] "Bezüglich strömungstechnischer Fragen wollen wir natürlich auch die Versuche mit glatten und rauhen Rinnen auf große Reynoldssche Zahlen erweitern. Ich habe zwei große Kessel zu diesem Zwecke aufgebaut und einen Wasserturm errichtet. Ich glaube aber, dass für diese grundlegenden Versuche eine Doppelarbeit nicht schädlich ist, insbesondere, weil man doch die Versuche als Anleitung für Ausarbeitung der Theorie benutzen möchte." Kármán to Prandtl, 2 August 1928. MPGA, Abt. III, Rep. 61, Nr. 792.

[9] "Ich habe ausserdem ausgerechnet, was man über das Widerstandsgesetz aussagen kann, wenn man folgende Annahmen macht: (a) Das Geschwindigkeitsprofil in ähnlichen Röhren ist nur von der Spannung an der Wand abhängig. (b) Der Mischungsweg im Prandtl'schen Sinne ist in der Nähe der Wand mit der Wandentfernung proportional. (c) Beim Übergang zur Laminarschicht schreibt man in dem Prandtl'schen Ansatz statt τ die Grösse $\tau - \mu \frac{dU}{dn}$. Mit diesen Ansätzen kommt man zu dem

Kármán elaborated this approach in a paper on "Mechanical Similarity and Turbulence" published in the Proceedings of the Göttingen Academy of Science. His similarity hypothesis consisted in the assumption that the mean turbulent fluctuations at different points are similar and differ only with respect to length and time scale. He derived from this assumption the same expressions for the velocity fluctuations ($l\frac{dU}{dy}$) and the turbulent stress ($\rho l^2(\frac{dU}{dy})^2$), but with a different interpretation of the characteristic length. "Our similarity approach provides much more," Kármán distinguished his mixing length concept from that of Prandtl, "insofar as it contains an equation for the determination of the mixing length, which is in first approximation the ratio of the two first differential quotients of the main flow velocity." Kármán's equation for the mixing length was

$$l = k\frac{\frac{dU}{dy}}{\frac{d^2U}{dy^2}}$$

with k as "the only constant that enters the here developed theory."[10]

In his Göttingen paper Kármán applied this approach to turbulent channel and pipe flow. For the velocity profile he obtained a logarithmic formula instead of the 1/7th law, and for the resistance coefficient a formula that he had already mentioned in his letter to Burgers in December 1929. In August 1930 he used the opportunity of the Third International Congress for Applied Mechanics in Stockholm for another presentation of his approach with particular emphasis on the turbulent skin friction of a flat plate (von Kármán, 1931).

Prandtl duly acknowledged Kármán's achievement in the forthcoming communications of his laboratory, but he claimed that he had arrived at the same results (Fig. 6.2) "at a time when Kármán's papers had not yet been known, so that once more, like ten years ago with the same problem, the thoughts in Aachen and Göttingen followed parallel paths".[11] In Kármán's recollection, however, Prandtl was "crestfallen" and "chagrined". The quest for a "universal law of turbulence" had turned into a "first-class rivalry" (von Kármán, 1967, pp. 135–138).

According to the dates of Kármán's and Prandtl's publications it is obvious that Kármán was the winner in this race. As we learn from their contemporary correspondence, however, the rivalry involved other concerns. The bone of contention became a report in the journal *Werft, Reederei, Hafen* about a conference at which Prandtl was

---

Widerstandsgesetz: [...] Die einzige wesentliche Konstante ist dabei der Proportionalitätsfaktor des Mischungsweges in der Nähe der Wand." Kármán to Burgers, 12 December 1929. TKC 4.22.

[10] "Nun liefert aber unsere ähnlichkeitsbetrachtung wesentlich mehr, indem wir eine Bestimmungsgleichung für die Mischlänge erhalten, die in erster Näherung als Verhältnis der zwei ersten Differentialquotienten erscheint... [k] ist die einzige Konstante, die in die hier entwickelte Theorie der Turbulenz eingeht." (von Kármán, 1930, pp. 64–65).

[11] "... sind also beide Darstellungen miteinander identisch. Es mag dabei erwähnt werden, dass unsere Darstellung aus einer Zeit stammt, in der die Kármánschen Arbeiten noch nicht bekannt waren, so dass hier wieder einmal, wie schon vor 10 Jahren bei demselben Problem, die Gedankengänge in Aachen und Göttingen ziemlich parallel gelaufen sind." (Prandtl, 1932, p. 21).

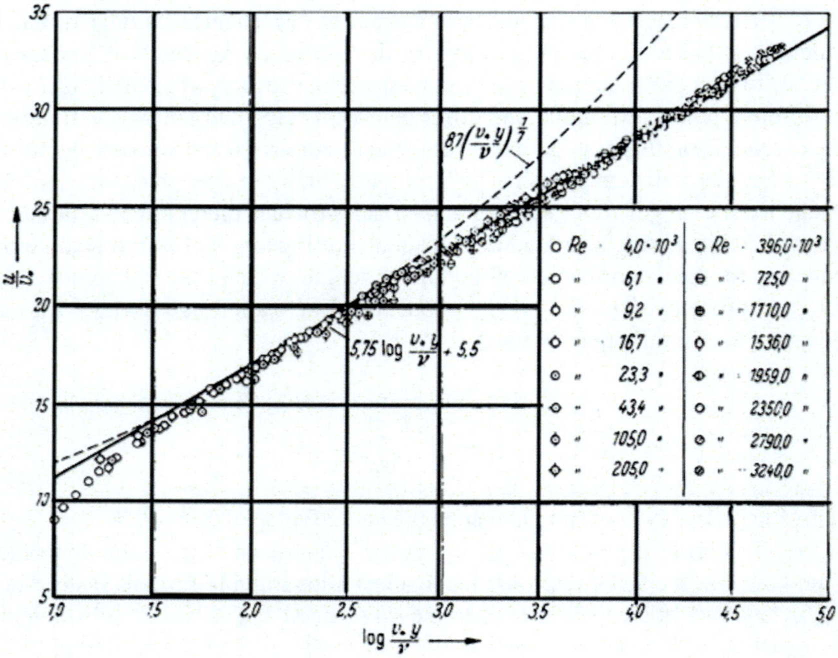

Fig. 6.2 Prandtl's presentation of the logarithmic wall law for the turbulent velocity distribution (made dimensionless with the "friction velocity" $v_\star = \sqrt{\tau/\varrho}$) (Prandtl, 1932, pp. 20–22, 1933, p. 110)

given the primacy. In a normal priority dispute it would have been sufficient for Kármán to point to his publications from the year 1930 in order to correct this erroneous view, but "who reads G. N. [Göttinger Nachrichten] and Stockholm Kongress," he complained in a letter to Prandtl; he wanted that his priority is acknowledged in the communications of Prandtl's Göttingen laboratory which were widely regarded as the "standard for practitioners."[12] In other words: the quest for a universal theory of turbulence was not just a competition for scientific laurels—it had become a concern for practical engineering as well.

The "log law", as the logarithmic distribution of turbulent velocities close to the wall and the related skin friction formulae, both for smooth pipes and flat plates, became famous among turbulence researchers, represented a major theoretical achievement. It demonstrated like few other theoretical results the pertinence of the mixing length concept for practical concerns. In 1933 Prandtl devoted a comprehensive review in the *Zeitschrift des Vereins Deutscher Ingenieure* to these accomplishments "that are also of immediate interest for the practitioner" ("die auch für den

[12] "Dabei beachten Sie, dass Göttinger Ergebnisse ein Standardwerk für Praktiker sind, so dass wenn doch die Sachlage einseitig dargestellt ist, bleibt meine Rolle in der Sache ewig begraben. Wer liest G. N. und Stockholmer Kongress?" Kármán to Prandtl, 26 September 1932. AMPG, Abt. III, Rep. 61, Nr. 793. For more detail see Eckert (2017, Sect. 6.8).

Praktiker von unmittelbarem Interesse sind") (Prandtl, 1933, p. 105). Kármán, who moved by this time permanently to Pasadena as director of the Guggenheim Aeronautical Laboratory at the California Institute of Technology (GALCIT), reviewed the recent results on "Turbulence and Skin Friction" for the *Journal of the Aeronautical Sciences* (von Kármán, 1934). Both included extensive comparisons of theory and experiment from naval and aeronautical laboratories. Despite broad agreement about the final formulae, however, Prandtl's and Kármán's approaches differed with regard to the meaning of the mixing length and the fundamental constants of the resulting "log law". Nor did their rivalry for a theory of turbulence end there. And they were not the only ones in this quest.

6.4 Taylor's Vorticity Transport Concept

"Some years ago, at a time when I was ignorant of the work of Prandtl, I put forward a somewhat similar theory," Taylor remarked in 1932 in response to Prandtl's mixing length concept, "but it differed in one very significant feature. I supposed that each particle of fluid retained the vorticity, but not the momentum, of the layer from which it started." In this way Taylor referred to his paper from 1915 on "Eddy motion in the atmosphere" in which he had regarded the turbulent mixing "as an effect of diffusion of vorticity rather than as a diffusion of momentum [...] though so briefly that it appears to have escaped notice." But Taylor did not content himself with a warm-up of his older concepts. In addition to the mixing of fluid particle velocities he analysed the diffusion of heat. According to Prandtl's momentum transport concept there should be no difference in the mixing of temperature and velocity in the wake of a heated rod, whereas Taylor obtained from his vorticity transport theory different distributions. "Measurements made at the National Physical Laboratory show a large difference between the distributions of temperature and velocity", Taylor referred to recent experimental corroboration of his view (Taylor, 1932, pp. 686, 701–702).

"In the last weeks I studied your old papers of 1915 and 1922 with greatest interest," Prandtl responded, "I think that if I had known these papers, I would have found the way to the turbulence earlier." His enthusiasm was genuine. "In the meantime we have studied experimentally the flow of a cold jet of air through a warm room, and found, that also in this case the turbulence is of your kind," he informed Taylor about another experimental corroboration. He concluded that Taylor's theory revealed

> ...a very important new fact concerning turbulence. It demonstrates, that there are two different forms of turbulence, one belonging to the fluid motions along the walls and the other belonging to the mixture of free jets. In the first the principal axis of vorticity is parallel to the direction of the main flow, in the other this direction is perpendicular to the flow.[13]

---

[13] Prandtl to Taylor, 25 July 1932. MPGA, Abt. III, Rep. 61, Nr. 1653. (The letter is written in English.)

Unlike Taylor's collaborator at the National Physical Laboratory who contrasted "Prandtl's theory" against "Taylor's theory", Taylor himself did not perceive his theory as a replacement of Prandtl's mixing length approach. He rather distinguished different fields of application. "In one case the conditions realised in practice seem to be rather close to one particular case and in the other case to the other."[14]

Prandtl also acknowledged the virtues of Taylor's vorticity transport theory in his report on recent advances in turbulence research:

> The turbulent velocity exchange is quite closely related to the heat exchange. [...] For the processes in the wake flow behind moving bodies, however, Taylor has recently shown that here the heat exchange is twice as strong as the exchange of velocity, so that the temperature and velocity curves have noticeably different shapes. [...] Unpublished experiments in Göttingen by P. Ruden [1903–1970] showed that Taylor's law of exchange also applies to the spreading of jets.
>
> It follows, then, that on closer inspection one has to distinguish between two types of turbulence, which differ in their mechanism. One can be called "wall turbulence" and the other "jet turbulence".[15]

In the published report of his experiments Ruden concluded that an extension of Prandtl's theory would require the consideration of rotational motion in addition to the translational movements that effected the momentum transport. However, that was left for the future. "To sum up, even with the relatively simple problem of free turbulence, there are still many open questions whose answers will be urgently needed for a better understanding of turbulent flows."[16]

For the time being, Prandtl and Taylor did not attempt to extend the mixing length concept. Nor did Kármán further invest much efforts to this end. The attention shifted from the mixing length concept towards direct measurements of turbulent velocity fluctuations from which statistical quantities such as correlation functions and the spectrum of turbulence could be derived. This opened the gateway towards a statistical theory of turbulence.

---

[14] Taylor to Prandtl, 14 January 1933. MPGA, Abt. III, Rep. 61, Nr. 1653.

[15] "Mit dem turbulenten Geschwindigkeitsaustausch steht in ziemlich engen Beziehungen der Wärmeaustausch. [...] Für die Vorgänge bei der Nachlaufströmung hinter bewegten Körpern hat jedoch kürzlich Taylor gezeigt, daß hier der Wärmeaustausch doppelt so stark ist wie der Austausch der Geschwindigkeit, so daß also die Temperatur- und die Geschwindigkeitskurve merklich verschiedene Gestalt haben. [...] Unveröffentlichte Göttinger Versuche von P. Ruden zeigten, daß auch bei der Strahlausbreitung das Taylorsche Austauschgesetz herrscht. Es ergibt sich also, daß man bei genauerem Zusehen zwei Arten der Turbulenz zu unterscheiden hat, die sich in ihrem Mechanismus unterscheiden. Man wird die eine 'Wandturbulenz' und die andere 'Strahlturbulenz' nennen können." (Prandtl, 1933, pp. 842–843).

[16] "Man muss nur neben den bisher allein in Betracht gezogenen translatorischen Bewegungen der einzelnen Flüssigkeitsballen auch eine rotatorische zulassen. [...] Zusammenfassend ist zu sagen, dass selbst bei dem verhältnismässig einfachen Problem der freien Turbulenz immer noch viele Fragen offen stehen, deren Beantwortung für ein besseres Verständnis der trubulenten Strömungen dringend nötig sein wird." (Ruden, 1933, pp. 377–378).

Chapter 7
The Statistical Theory of Turbulence

Abstract The statistical theory of turbulence, perceived as a theory of correlation functions between fluctuating quantities, began in the early 1920s with Taylor's analysis of the "diffusing power" of turbulence and reached a climax in the 1930s when hot-wire measurements of grid-generated wind tunnel turbulence enabled a comparison of theory with experimental data. The statistical concept at this stage was mainly developed by Taylor in Cambridge and von Kármán in Pasadena; the crucial measurements were made in wind tunnels in England (Simmons, Salter), Germany (Prandtl, Reichardt) and the USA (Dryden and collaborators). In 1939, Kolmogorov started another research effort with the focus on the smallest scale of turbulence for which he postulated universal features. This theory was published in 1941, but remained unknown outside Russia until the end of the Second World War.

7.1 Correlation Functions as Characteristics of Turbulent Flow

"The theory of turbulence by its very nature cannot be other than statistical," an authoritative textbook argued about the conceptual nature of theories of turbulence (Monin and Yaglom, 1971, pp. 3–4). Strictly speaking, the statistical theory of turbulence began with the treatise of Boussinesq (1877) that accounted for turbulent motion in terms of equations for mean values and an eddy viscosity that varied somehow according to the specific case. But Boussinesq's approximation ignored terms that involved averages of fluctuating quantities and thus could not be elaborated into a full-fledged statistical theory of turbulence. The more comprehensive averaging of the Navier–Stokes equations lead Reynolds in 1895 to the formulation of a set of equations that appeared better suited for statistical elaboration—but elusive with regard to experimental test. The mixing length concept may be regarded in the same vein as statistical: it adopted from Reynolds's equations the stress term and merged it with Boussinesq's eddy viscosity so that it could be adjusted to specific flows. For this reason it was rated as a semi-empirical theory which deserved further attention for many practical application, but it did not exhaust the need for a statistical theory of turbulence.

M. Eckert, *Turbulence—an Odyssey*, History of Physics,
https://doi.org/10.1007/978-3-030-91459-2_7

The remarks about correlation functions in the correspondence between Prandtl and Taylor in the early 1930s hint at the roots from which Taylor developed a few years later a statistical theory of turbulence that truly deserved this name. He had made the first step in this direction ten years earlier in the context of his investigations on atmospheric turbulence. With reference to his paper on eddy motion in the atmosphere from 1915 (see Sect. 4.3) he pointed out that turbulence mixed "heat and other diffusible properties" in the same way as "molecular agitation gives rise to molecular diffusion", but whereas the latter could be dealt with in terms of the kinetic theory of gases, the "diffusing power" of turbulence was unknown. He assumed that turbulent motion was uniformly distributed throughout space and considered the turbulent diffusion of heat in one direction (Taylor, 1922, p. 196).

Taylor adopted for this analysis the so-called Lagrangian approach, where the motion of a particle is followed from its initial coordinates (a, b, c) at time $t = 0$ to some later position (x, y, z) at time t. "If a physicist were to try to define the characteristic features of any particular case of turbulent motion, with a view to discussing statistically its effect as a conductor of heat, he would probably first fix his attention on the mean energy of the motion." In this way Taylor focused his investigation on the first characteristic of a statistical theory, the mean energy, including its change in time. The problem was similar to a "drunkard's walk" except that now mean values of continuously varying quantities were involved. Whereas in the discrete case of the drunkard's walk the correlation of subsequent steps could be set 0, the continuous turbulent motion had to assume a non-zero correlation between the movement from one time to the next. Taylor derived a formula for the mean-square distance X traversed by a particle in time T (Taylor, 1922, p. 207)

$$\langle X^2 \rangle = 2\langle u^2 \rangle \int_0^T \int_0^t R_\xi \, d\xi \, dt$$

where u is the velocity of the particle and R_ξ is the correlation function between the velocity of the particle at one instant and that at a time ξ later:

$$R_\xi = \frac{\langle u_t u_\xi \rangle}{\langle u^2 \rangle}.$$

Thus the correlation function R_ξ was exposed as the critical statistical quantity for a theory of turbulent diffusion. It also suggested a measure for the scale of turbulence that was conceptually different from the mixing length because it did not depend on assumptions about the turbulent exchange of momentum or vorticity. Unlike the mixing length approach, the statistical approach via the correlation function implied a continual change of the physical property under consideration. As Taylor recalled many years later,

> I was never satisfied with the mixture-length theory, because the idea that a fluid mass would go a certain distance unchanged and then deliver up its transferable property, and become identical with the mean condition at that point, is not a realistic picture of a physical process.[1]

Yet, the correlation function appeared as elusive as other quantities of turbulent flow as long as the fluctuating velocities escaped direct measurement. This changed with the application of the hot-wire technique to air flow measurements in wind tunnels.

7.2 Wind Tunnel Turbulence

Model testing in wind tunnels was based on the assumption that the coefficients of lift and drag measured in a wind tunnel and those in free flight are the same if the model is geometrically similar and the Reynolds number is the same. But since Eiffel's and Prandtl's experiments with spheres in the wind tunnel it was clear that turbulence affected model testing (see Sect. 3.5). Hugh Dryden (1898–1965), the director of the Aerodynamical Physics Section at the National Bureau of Standards in Washington, asked Prandtl in 1921 about "your idea as to the physical conception of the turbulence" because "the most important wind tunnel problem is a study of turbulence and its effects."[2]

Since the 1920s aerodynamic laboratories devoted a considerable research effort to wind tunnel turbulence. The degree of turbulence was modified by screens of different mesh size in a wind tunnel. In a study at the MIT it was found "that the turbulence tends to die out more rapidly downstream as the screen becomes finer" (LePage and Nichols, 1924, p. 5). But it was difficult to measure the degree of turbulence. In an effort to standardize wind tunnel tests engineers from the NACA compared the drag measurements on spheres obtained in various wind tunnels (Bacon and Reid, 1924). Wind tunnel turbulence cast doubt on the reliability of model testing. "The actual process of standardization still lies in the future" (Reid, 1925, p. 219). Dryden and his collaborators at the National Bureau of Standards performed a serious of measurements about the resistance of cylinders at different degrees of turbulence (effected by different wire meshes in the 54-in. tunnel of their laboratory) because "we can never be satisfied until we know more definitely the nature of turbulence; in other words, its cause and the exact manner in which it manifests itself as changes in the air speed, direction, and pressure" (Dryden and Heald, 1926, p. 465).

The decisive step towards this goal was the use of hot-wire anemometry for measuring the fluctuating velocity of the air stream in the wind tunnel. The principle was known for several years (see Sect. 5.5), but the involved electronics made it cumbersome to handle. The first hot-wire apparatus that was employed at the wind tunnel of the National Bureau of Standards was "very bulky, far from portable,

---

[1] Taylor (1970, p. 8), also quoted in Sreenivasan (2011, p. 137).

[2] Dryden to Prandtl, 5 March 1921. MPGA, Abt. III, Rep. 61, Nr. 361. For more detail see Eckert (2006, Sect. 5.5).

and in many respects inconvenient to use" (Dryden and Kuethe, 1929b, p. 381). But Dryden and his collaborators spared no effort to pioneer this technique for turbulence measurements, "perhaps only with great difficulty and complicated apparatus, but surely and with reasonable accuracy."[3]

It was not long before the hot-wire technique became also employed elsewhere for measuring wind tunnel turbulence. In December 1933 Taylor communicated to the Royal Society the results of an "Experimental Investigation and Analysis of the Velocity Variations in Turbulent Flow" conducted upon his request with a sophisticated hot-wire equipment at the NPL. The measurements were performed at variable distances in the turbulent wake behind a grid composed of horizontal and vertical strips that formed a mesh of one-inch square holes (Simmons and Salter, 1934). Similar measurements with hexagonal grids ("honeycombs") were conducted by Dryden's group with the goal to find means for reducing wind tunnel turbulence. "Moderate changes in the size, shape, and wall thickness of cells of the honeycomb were found to have little effect," Dryden concluded, but in each case "turbulence decreased with increasing distance between the honeycomb and the measuring station" (Dryden, 1931, p. 557). At Göttingen, too, the hot-wire technique was introduced. In June 1933 Prandtl's collaborator, Hans Reichardt, presented first results of hot-wire measurements in a special wind tunnel designed for studies of turbulent air flow near walls with a rectangular cross section of 25 cm by 100 cm (Reichardt, 1933). A year later, Reichardt was able to measure with two hot-wire probes directly the correlation functions of velocity fluctuations at neighboring points (Prandtl and Reichardt, 1934).

7.3 Taylor's Statistical Theory of Turbulence

By the 1930s, on the pillars of hot-wire measurements in wind tunnels, the stage was set for a statistical theory of turbulence. "I think that the purely descriptive side of turbulence is capable of considerable development," Taylor wrote to Prandtl in August 1932. He pointed to "the 'spectrum' of turbulence" which so far "has not received much attention. By means of a hot-wire and oscillograph one can (theoretically) obtain the frequency of occurrence of all turbulent velocities and can construct a frequency diagram."[4] Prandtl, too, raised great expectations into the hot-wire technique. He shared Taylor's opinion that the time was ripe for a statistical approach and informed him about Reichardt's hot-wire measurements in his laboratory from which "very important information can be obtained about the behaviour of turbulent flows."[5]

---

[3] Dryden to Prandtl, 8 May 1931. MPGA, Abt. III, Rep. 61, No. 361.

[4] Taylor to Prandtl, 18 August 1932. MPGA, Abt. III, Rep. 61, Nr. 1653.

[5] "Jedenfalls bin ich mit Ihnen einer Meinung darüber, dass durch das Studium dieser Korrelation sowie auch derjenigen Richtungsschwankungen und Größenschwankungen, für die wir auch eine Hitzdrahtanordnung vorbereitet haben, sehr wichtige Aufschlüsse über das Verhalten der turbulenten

The measurements of velocity fluctuations behind honeycombs (Dryden, 1931) and square meshes (Simmons and Salter, 1934), as well as Reichardt's and Prandtl's direct determination of correlation functions provided the experimental basis for Taylor's "Statistical Theory of Turbulence" published in a sequence of five successive papers in the *Proceedings of the Royal Society* from summer 1935 to Spring 1936 (Taylor, 1935a, b, c, d, 1936). He had outlined the main results of his theory in a letter to Prandtl already in March 1935. The pertinent passage from this letter deserves to be quoted at full length because it shows to what extent Taylor was committed to experiments on wind tunnel turbulence. The close association with experiment, in this case the measurements of decaying turbulence behind grids in the wind tunnel, is expressed by two formulae that he regarded as the main result of his theory:

> I have lately been doing a great deal of work on turbulence, basing my discussion on my work of "Diffusion by Continuous Movements" [see Sect. 7.1 and Taylor, 1922]. In the course of my work I have brought out two formulae which seem to have practical interest.
>
> The first concerns the rate of decay of turbulent energy in a wind stream. It is
>
> $$\frac{U}{\sqrt{\overline{u^2}}} = \frac{8}{A^2}\frac{x}{L} + \text{constant}.$$
>
> U = velocity of stream
> $\sqrt{\overline{u^2}}$ = longitudinal component of turbulent velocity
> x = distance down wind tunnel from any fixed point
> L = scale of disturbance. If the turbulence is produced by a grid [sketch of square mesh grid], $L = \frac{1}{2}$ mesh = $\frac{1}{2}D$. A is a constant which I calculated from Schlichting's measurements of the spreading of a wake in a stream which is non-turbulent outside the wake [see Schlichting, 1930]. It is approximately $A = 3.5$.
>
> I have compared this formula with some of Dryden's measurements behind a honeycomb and it seems to fit. It also fits [L. F. G.] Simmons' measurements made with turbulence of a very different scale.
>
> The formula is only true if the "Reynolds number of turbulence", namely $\frac{\sqrt{\overline{u^2}}L}{\nu}$, is greater than, say, 100 which is the condition where the ordinary viscous decay of a pattern of a scale L begins to give a rate of decay comparable with that given by my formula. The rate of decay then becomes a logarithmic decrement and the change of $\frac{U}{\sqrt{\overline{u^2}}}$ ceases to be linear down the channel.
>
> Have you made any measurements in Göttingen which bear on this point?
>
> The other formula which comes out of the same theory is concerned with the effect of turbulence on the boundary layer. If you assume that a sphere has a given drag coefficient (in the middle of the critical region), say 0.3, when the transition between laminar and turbulent boundary layer occurs on a certain zone (say at 45° from the front stagnation point) then my theory gives the result that the Reynolds number, R_{crit}, corresponding with that condition, is
>
> $$\frac{Ua}{\nu} = R_{\text{crit}} = f\left\{\frac{\sqrt{\overline{u^2}}}{U}\left(\frac{a}{L}\right)^{\frac{1}{5}}\right\}$$

Strömungen erzielt werden können." Prandtl to Taylor, 23 December 1932. MPGA, Abt. III, Rep. 61, Nr. 1653.

where f is a function which is not determined in the present state of the theory. a is the radius of the sphere and L is the scale of the turbulence.

Have you made any sphere tests varying the size of the sphere and keeping L constant, i.e. keeping the honeycomb or turbulence screen constant?[6]

In his theory on turbulent diffusion fifteen years ago Taylor had adopted the Lagrangian conception. From this perspective a characteristic turbulent length scale could be defined in terms of a correlation function R_ξ of fluctuations separated by a time increment ξ. In order to enable comparison with hot wire measurements in wind tunnels Taylor now adopted also the Eulerian conception and defined another characteristic length L in terms of a correlation function R_y between simultaneous velocity fluctuations u_0 and u_y separated within a cross section of the air stream by a lateral spatial distance y:

$$L = \int_0^\infty R(y)dy \quad \text{with} \quad R(y) = \frac{\overline{u_0 u_y}}{\sqrt{\overline{u_0^2}}\sqrt{\overline{u_y^2}}}$$

The formula on the decay of turbulent energy quoted in the letter to Prandtl was limited to isotropic flow (where the average values of the velocity fluctuations are the same in all directions) which was nearly fulfilled for grid-generated turbulence in wind tunnels at some distance from the grid. The formula resulted from equating the rate of loss of kinetic energy of the turbulence per unit volume

$$-\frac{3}{2}\rho U \frac{d}{dx}\overline{u^2}$$

with the rate of dissipation for which Taylor derived

$$15\mu \frac{\overline{u^2}}{\lambda^2}$$

where λ is a "microscale" of turbulence defined by the curvature of R_y. Taylor regarded λ "as a measure of the diameters of the smallest eddies which are responsible for dissipation of energy" (Taylor, 1935a, p. 437). As a consequence that could be tested by hot-wire measurements of grid-generated turbulence in wind tunnels he derived the linear law for the decay of turbulence with increasing distance x from the grid:

$$\frac{U}{u'} = \frac{5x}{A^2 M} + \text{constant}$$

where u' is the measured velocity fluctuation, M the mesh size and A "an absolute constant for all grids of a definite type, e.g., for all square-mesh grids or honeycombs"

[6] Taylor to Prandtl, 2 March 1935. MPGA, Abt. III, Rep. 61, Nr. 1654. For a detailed analysis of Taylor's concept with the hindsight of subsequent developments see Sreenivasan (2011, Sect. 4.5).

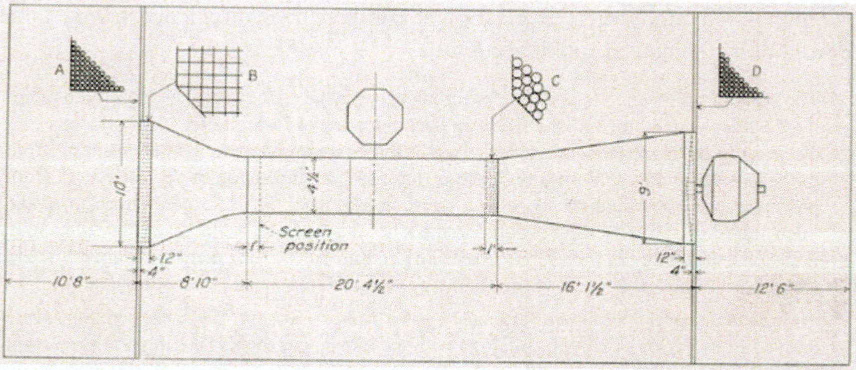

Fig. 7.1 Diagram of the wind tunnel used by Dryden and his collaborators at the National Bureau of Standards for hot-wire measurements of turbulent velocity fluctuations (Dryden et al., 1936, p. 3)

(Taylor, 1935a, p. 440). In his letter to Prandtl he had already communicated this linear law in a slightly different form.

Taylor regarded the applicability of his theory to wind tunnel experiments as a major asset. One of his subsequent papers was devoted to the comparison with Prandtl's and Reichardt's measurements of the lateral distribution of turbulent energy (Taylor, 1935c), another (Taylor, 1935d) to the diffusion of heat in the wake of a heated source in the turbulent air stream of a wind tunnel (Schubauer, 1935). A year later, Dryden and his collaborators published the results of a comprehensive comparison of Taylor's formulae with hot-wire measurements in the wind tunnel of the National Bureau of Standards. They used a special hot-wire probe in which the distance between two wires could be varied for a direct measurement of the correlation function. The degree of turbulence was varied by inserting screens with different meshes in the air stream (Fig. 7.1).

By and large, the hot-wire measurements at the National Bureau of Standards confirmed both Taylor's formulae for the decay of grid-generated turbulence and for the critical Reynolds number of spheres (Dryden et al., 1936, p. 29). The main lesson which Dryden drew from these experiments was that wind tunnel turbulence could be characterized by two statistical quantities, the intensity expressed by $\frac{\sqrt{\overline{u^2}}}{U}$ and the scale of turbulence expressed by $L = \int_0^\infty R(y)dy$, which were both amenable to hot-wire measurement. "Taylor's theory represents a movement from the 'atomic' or discontinuous to a continuous representation of turbulence and again emphasizes the desirability of statistical treatment," Dryden concluded in a review presented on January 27, 1937, at the Fifth Annual Meeting of the American Institute of the Aeronautical Sciences (IAS). "Progress will probably continue in this direction" (Dryden, 1937, p. 274). As a first step in this direction Dryden derived from Taylor's theory slightly different formulae that could be better compared with hot-wire measurements and offered an interpretation of grid-generated turbulence in wind tunnels devoid of mathematical sophistication:

> The body of fluid which is in a state of turbulence may be conceived to be filled with rather large eddies which move to and fro in an irregular manner losing energy by collisions [...] The energy of this macro-turbulence is redistributed among smaller eddies resulting from the collisions, and this micro-turbulence is dissipated by the action of viscosity [...]. Both processes occur simultaneously with a quasi-steady state. The rate of dissipation of the micro-turbulence by viscosity is equal to the rate of dissipation of the macro-turbulence by collisions, the energy in both forms decreasing continuously. The energy per unit volume in the micro-turbulence is equal to the energy per unit volume of the macro-turbulence. [...]
>
> The rate of decay and the change of scale as given by Eqs. (18) and (19) [these equations were derived from Taylor's theory by analogy with the kinetic gas theory] are in good agreement with the experimental results obtained at the National Bureau of standards. (Dryden, 1937, pp. 278–279)

7.4 The Kármán–Howarth Theory

Kármán announced at the same IAS-meeting a theory that he was elaborating together with Leslie Howarth (1911–2001), a research fellow from the University of Cambridge who had come to Caltech in the summer of 1936 for a one-year sojourn. Kármán introduced his presentation with a praise of Taylor's statistical theory which "raised considerable interest, because it is concerned with the important problem of wind-tunnel turbulence and its results could be compared directly with experimental work done by Dryden in this country and by Fage, Townend and Simmons in England."[7] Kármán started from the "correlation tensor" made up from correlation coefficients of fluctuating velocities at two points separated by a distance r, and aimed at a general theory of isotropic turbulence that included Taylor's theory as a special case (von Kármán, 1937a, p. 131). A hunch of the general theory may be obtained from the summary with which Kármán and Howarth introduced their extensive publication:

> The correlation coefficients between two arbitrary velocity components at two arbitrary points form a tensor called the correlation tensor. Due to isotropy the tensor corresponding to one fixed and one variable point has spherical symmetry; due to the condition of continuity it is completely determined by one scalar function. The mean products of the derivatives of the velocity fluctuations are expressed by the derivatives of the tensor components; in this way laborious calculations for obtaining such mean values are eliminated. The correlation between three components (triple correlation) is discussed. After developing the kinematics of isotropic turbulence the dynamical problem of the change of the various mean values with time is considered. It is shown that using the equations of motion a partial differential equation connecting the double and triple correlation functions can be established. (von Kármán and Howarth, 1938, p. 192)

[7] Arthur Fage (1890–1977), Herbert Charles Henry Townend (1896–1943) and L. F. G. Simmons were employed at the National Physical Laboratory and collaborated with G. I. Taylor.

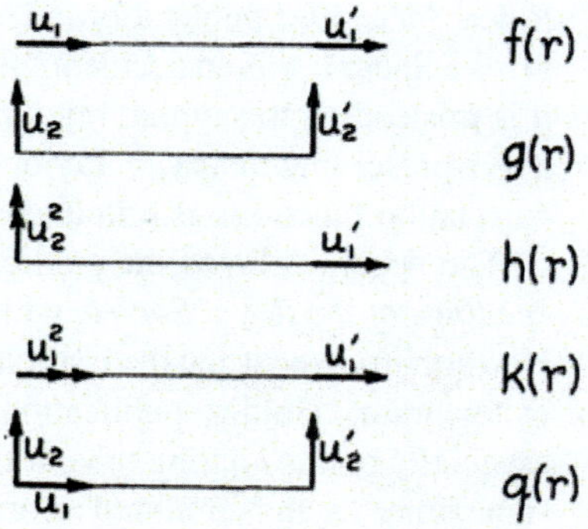

Fig. 7.2 The double- and triple-correlations functions for which Kármán and Howarth derived equations for isotropic turbulence (von Kármán and Howarth, 1938, p. 202)

The theory involved five double- and triple-correlation functions (Fig. 7.2). The analysis amounted to an equation which Kármán and Howarth called "the fundamental equation for the propagation of the correlation function $f(r)$":

$$\frac{\partial(f\overline{u^2})}{\partial t} + 2(\overline{u^2})^{3/2}\left(\frac{\partial h}{\partial r} + \frac{4}{r}h\right) = 2\nu\overline{u^2}\left(\frac{\partial^2 f}{\partial r^2} + \frac{4}{r}\frac{\partial f}{\partial r}\right),$$

where r is the distance between the two points, $f(r,t)$ and $h(r,t)$ are correlation functions as described in Fig. 7.2, and $\overline{u^2}$ is the mean square turbulent velocity (which is the same in all directions in the case of isotropic turbulence). The "Kármán–Howarth equation", as it became famous, describes the temporal evolution of turbulence in terms of two-point correlations. When $h(r,t)$ was set zero (i.e. triple-correlations neglected) the solution of this equation yielded the same decay law for $\overline{u^2}$ (i.e. the mean kinetic energy) which Taylor had derived for the decaying turbulence behind a grid in a wind tunnel (von Kármán and Howarth, 1938, p. 207):

$$\frac{1}{U}\frac{d}{dx}\overline{u^2} = -10\nu\frac{\overline{u^2}}{\lambda^2}$$

where U is the velocity of the air stream in the wind tunnel and λ Taylor's micro-scale (derived from the curvature of the transverse correlation function $g(r)$). Furthermore, Kármán and Howarth invoked the feature of "self-preserving" of their correlation functions, i.e. that "the form of these curves is the same at all instants although the actual length scale varies". From this similarity principle they derived a general law for the decay of turbulence at some distance x behind a grid which amounted to Taylor's linear law only if a parameter was given a special value (von Kármán and Howarth, 1938, p. 211).

The underlying physical assumptions of the Kármán–Howarth theory, however, met with Taylor's objections. Taylor denied that his theory represented merely a special case. The bone of contention was the linear law for the decaying turbulence behind a grid. Kármán had sent Taylor a first version of his and Howarth's theory already before the IAS-meeting with the request to present it to the *Proceedings of the*

Royal Society for publication. "The fact that you find a self-preserving law that also gives a linear law is an accident," Taylor remarked about Kármán's derivation. "Thus it is misleading to say that my non-self preserving assumption is a limiting case of a self-preserving theory."[8] Taylor regarded Kármán's assumption of self-preserving correlation functions as unfounded.

The debate delayed the publication of the Kármán–Howarth theory in the *Proceedings of the Royal Society* which Kármán had requested in January 1937. Leslie Howarth, too, regarded the issue still unsettled when he asked Kármán in April 1937 by telegram: "Hold up publication of joint Royal paper. Essential changes necessary, especially points relating to your theory." (Leonard and Peters, 2011, p. 116). By the same time, Taylor submitted an article to the *Journal of the Aeronautical Sciences* in response to Kármán's paper to the IAS-meeting in which he verified a relationship between correlation functions that Kármán had presented as an asset of his theory—but questioned its underlying prerequisites. ("It will be seen that the observed $R_1(r)$ curve is very different indeed from that calculated on the assumption that turbulence is 'self-preserving' and that the $R_1(r)$ curve preserves its shape.") With regard to the involved physical mechanism of the decay of turbulence, Taylor regarded Kármán's view as fundamentally different from his own:

> It seems that the essential contrast between von Kármán's and the author's assumptions lies in the difference between their views as to the significance of the "inertia" terms in the mechanism of turbulent motion. Von Kármán considers that they have no effect, at any rate in the problems with which he is concerned. In the author's view they represent the fundamental mechanical cause which controls the dissipation of energy in turbulent motion.
>
> The idea the author has in mind is that the turbulence in a wind tunnel is found to be diffusive, i.e., the average distance between pairs of particles A and B continually increases. Suppose that a random vorticity is set up in a fluid and that this gives rise to turbulent motion which proves to be diffusive. Choose particles A and B a short distance d_0 apart on a vortex line where the initial vorticity is ω_0. If the viscosity is very small the vortex line will remain a vortex line and the vorticity will change in accordance with the law $\omega = d^2\omega_0^2/d_0^2$ where d is the distance apart at any time. Since the motion is diffusive, the average value of d^2/d_0^2 will continually increase so that the average value of ω^2 will continually increase. This process may be expected to continue until ω becomes so large that the vorticity is destroyed by viscosity faster than it is produced by the stretching of vortex filaments. This process can only be represented by the "inertia" terms in the equation of motion. (Taylor, 1937, p. 315)

Taylor had analysed this process in more detail in a paper on the "Mechanism of the Production of Small Eddies from Large Ones" co-authored with his doctoral student Albert Edward Green (1912–1999). "It appears that nothing but a complete solution of the equations of motion in some special case will suffice to illustrate the process of grinding down of large eddies into smaller ones" (Taylor and Green, 1937, p. 502). Thus they derived for a special (non-turbulent) flow at time $t = 0$ the subsequent motion from which they determined the mean rate of dissipation of energy in powers of t. They found that the rate of dissipation of energy increases

[8] Taylor to Kármán, 24 January 1937. Quoted in Leonard and Peters (2011, p. 116). For a detailed discussion of the exchange between Taylor and Kármán on the issue of "self-preservation" and its further development see Leonard and Peters (2011, Sect. 3.3) and Sreenivasan (2011, Sect. 4.6.3).

due to a local increase of vorticity until it reaches a maximum and then begins to decay. Taylor concluded in a letter to Kármán that "this puts the whole idea of my dissipation theory of turbulence on a sounder basis" (quoted in Sreenivasan, 2011, p. 161).

The publication of the Kármán–Howarth theory was delayed for several months. In September 1937 Kármán added a final section which mirrors Taylor's objections. Here Kármán derived for the limiting case of large Reynolds numbers under the assumption of self-preservation ("that the correlation functions preserve their shape and only their scale is changing") two equations (von Kármán and Howarth, 1938, Eqs. 75 and 76) which he perceived as the most general mathematical expression for the decay of turbulence behind a grid. "It appears that the next step in the development of the theory should be to find the physical mechanism which is behind the mathematical relations (75) and (76)," he concluded this section (von Kármán and Howarth, 1938, p. 215).

Meanwhile Taylor further investigated what he regarded as the essence of turbulence: the role of vortex-stretching. In October 1937 he submitted to the Royal Society a paper on the "Production and Dissipation of Vorticity in a Turbulent Fluid". Once more he referred to hot-wire measurements at the National Physical Laboratory which proved that vorticity is produced by vortex filaments "three times as fast as it is disappearing". Taylor concluded "that the stretching of vortex filaments must be regarded as the principal mechanical cause of the high rate of dissipation which is associated with turbulent motion" (Taylor, 1938a, p. 23).

In a late sequel to his statistical theory Taylor focused on the spectrum of turbulence as another property that was accessible to direct hot-wire measurements in wind tunnels. He assumed that for large velocities U of the main air stream the turbulent velocity u is swept largely unchanged along x in the direction of the main flow—an assumption that became known as frozen-flow hypothesis. From this assumption he derived the spectrum of turbulence as the Fourier transform of the correlation function R_x (in mathematics this is known as the Wiener–Khintchine theorem). The spectrum shed light on the distribution of small and large scale components of turbulence. It had also a bearing on the dissipation of turbulence, as Taylor remarked at the end of this analysis. The increase of high frequency components with the speed of the air stream "seems to confirm the view frequently put forward by the author that the dissipation of energy is due chiefly to the formation of very small regions where the vorticity is very high." Yet the spectrum also confirmed the similarity views: "Apart from these very small regions the turbulence behind a grid is similar at all speeds," Taylor added in conclusion (Taylor, 1938b, p. 489).

7.5 K41

By the same time, the Russian mathematician Andrey Nikolaevich Kolmogorov (1903–1987) developed a keen interest in turbulence. It is not entirely clear what prompted Kolmogorov's effort (Falkovich, 2011, p. 214). Perhaps he regarded the statistical theory of turbulence as a target of opportunity to extend his preceding work on stochastic processes? Or it resulted from his recent election as member of the USSR Academy of Science and the appointment as secretary of the section of Physics and Mathematics of the Institute of Theoretical Geophysics? In any case it was not a solitary effort. As Kolmogorov recalled many years later (quoted in Yaglom, 1994, p. 2),

> it soon also became clear to me that there was no chance to develop a closed purely mathematical theory. For lack of such a theory it was necessary to use some hypotheses based on the results of treatment of experimental data. Therefore it was important to find talented collaborators able to work in such a mixed regime who could combine theoretical studies with the analysis of experimental results. In this respect I was quite successful.

The collaborators recruited for this quest were Alexander Mikhailovich Obukhov (1918–1989), Mikhail Dmitrievich Millionshchikov (1913–1973), Andrei Sergeevich Monin (1921–2007) and Akiva Moiseevich Yaglom (1921–2007), all of them students by this time. "The study of Taylor's paper of 1935 on isotropic turbulence was especially inspiring for AN and led him to ideas about further study of such turbulence," Yaglom recalled about the sources from which they started their investigation (Yaglom, 1994, p. 3). Kolmogorov's theory "undoubtedly marked the greatest achievement in the theory of turbulence since the time of Reynolds." This is how Yaglom rated in retrospect its impact for the further development of the statistical theory of turbulence. It was "basic to all subsequent developments in the field during the second half of this century" (Yaglom, 1994, p. 8). The theory became famous among turbulence researchers as "K41".[9] Despite its mathematical outlook the theory was based on empirical observations of turbulent flows. One of Kolmogorov's later students, Yakov Grigorevich Sinai (born 1935), recalled that Kolmogorov arrived at his results after "half a year analyzing experimental data" (cited in Falkovich, 2011, p. 217).

The combined effort of Kolmogorov and his young collaborators started in 1939 with a paper by Millionshchikov on the "Degeneration of homogeneous isotropic turbulence in viscous incompressible fluids", followed by a paper which Kolmogorov submitted in December 1940 to the Soviet Academy of Science on "The Local Structure of Turbulence in Incompressible Viscous Fluid for Very Large Reynolds Numbers". The gist of Kolmogorov's concept was the assumption that at sufficiently high Reynolds numbers any turbulent flow becomes "locally isotropic" in a sufficiently small domain. The physics behind this concept was referred to a footnote and reads as follows[10]:

> For very large R the turbulent flow may be thought of in the following way: on the averaged flow [...] are superposed the 'pulsations of the first order' consisting in disorderly

[9] There are numerous accounts of Kolmogorov's theory. For a contemporary review see Batchelor (1947); for a review with hindsight of subsequent developments see Frisch (1995) and Falkovich (2011).

[10] Kolmogorov's papers were translated in English and published in the Proceedings of the Royal Society (Kolmogorov, 1991a, b) at the occasion of the 50th birthday of their first publication in the Proceedings of the USSR Academy of Science (Doklady Akademii Nauk SSSR). Another English translation was published in Friedlander and Topper (1961).

> displacements of separate fluid volumes, one with respect to another, of diameters of the order of magnitude $l^{(1)} = l$ (where l is the Prandtl's mixing path); the order of magnitude of velocities of these relative velocities we denote by $u^{(1)}$. The pulsations of the first order are for very large R in their turn unsteady, and on them are superposed the pulsations of the second order with mixing path $l^{(2)} < l^{(1)}$ and relative velocities $u^{(2)} < u^{(1)}$; such a process of successive refinement of turbulent pulsations may be carried until for the pulsations of some sufficiently large order n the Reynolds number
>
> $$R^{(n)} = \frac{l^{(n)}u^{(n)}}{\nu}$$
>
> becomes so small that the effect of viscosity on the pulsations of the order n finally prevents the formation of pulsations of the order $n + 1$.
>
> From the energetical point of view it is natural to imagine the process of turbulent mixing in the following way: the pulsations of the first order absorb the energy of the motion and pass it over successively to pulsations of higher orders. The energy of the finest pulsations is dispersed in the energy of heat due to viscosity.
>
> In virtue of the chaotical mechanisms of the translation of motion from the pulsations of lower orders to the pulsations of higher orders, it is natural to assume that in domains of the space, whose dimensions are small in comparison with $l^{(1)}$, the fine pulsations of the higher orders are subjected to approximately space-isotropic statistical regime. Within small time-intervals it is natural to consider this régime approximately steady even in the case, when the flow in the whole is not steady. (Kolmogorov, 1991b, p. 10, footnote 2)

Thus the concept of a cascade through which energy was transported step by step down to its dissipation at the smallest scale re-appeared in a new guise (without citation of Richardson who had arrived at this view from a different perspective, see Sect. 4.3). Yet it entailed more than Richardson's cascade view because of its statistical elaboration. Kolmogorov aimed at expressions for correlation functions like Taylor, Kármán and Howarth, but in contrast to their approach he focussed on correlations between velocity *increments*, i.e. *differences* of velocities at two points separated by a distance r. For their determination he made a "first hypothesis of similarity" which postulated for locally isotropic turbulence that the correlation functions depend only on the viscosity ν and the mean energy dissipation per unit mass ε. By dimensional analysis he derived from this requirement the smallest scale at which the kinetic energy of the smallest eddies was transformed in heat (Kolmogorov, 1991b, p. 12):

$$\eta = \frac{\nu^{3/4}}{\varepsilon^{1/4}}.$$

Kolmogorov called it "the scale of the finest pulsations, whose energy is directly dispersed into heat energy due to viscosity." In a "second hypothesis of similarity" he demanded that for the larger pulsations, i.e. for r large compared with η, the transfer of energy is independent from viscosity. This hypothesis enabled him to determine by dimensional analysis the double-correlation function ("second moment") of the longitudinal velocity increments:

$$B_{dd}(r) = \overline{(u'_d - u_d)^2} = C(\varepsilon r)^{2/3},$$

where u' and u are velocity components at two points P' and P separated by the distance r; the suffix d denotes the longitudinal component directed along $P'P$ (corresponding to the definition of $f(r)$ in the Kármán–Howarth-theory, see Fig. 7.2). C is "an absolute constant" (Kolmogorov, 1991b, p. 13).

In a sequel titled "Dissipation of energy in the locally isotropic turbulence", received on 30 April 1941 by the USSR Academy of Science, Kolmogorov dealt with the decay of turbulence in terms of the Navier–Stokes equations—without invoking the similarity hypotheses of his first paper. He started with "the known equation of Kármán for the isotropic turbulence in the sense of Taylor" (i.e. the Kármán–Howarth equation) and derived a corresponding differential equation for the second and third moments of the velocity increments, B_{dd} and B_{ddd}. He obtained the solution (Kolmogorov, 1991a, p. 15)

$$B_{ddd}(r) = \overline{(u'_d - u_d)^3} = -\frac{4}{5}\varepsilon r.$$

Next to this result on the third moment[11] Kolmogorov defined the "skewness"

$$S = \frac{B_{ddd}}{B_{dd}^{3/2}}$$

for which he regarded it "natural to assume" that it becomes constant for large r. That amounted to another derivation of the "two-thirds law" which was not based on the similarity hypotheses employed in his earlier paper:

$$B_{dd}(r) = C(\varepsilon r)^{2/3}, \quad \text{with} \quad C = (-4/5S)^{2/3}.$$

Finally Kolmogorov compared his results with data on the decay of turbulence behind a grid as published by Dryden and his collaborators (1936). After transforming the correlation functions for velocity increments into corresponding correlation functions for velocities, the comparison yielded a value for the constant in the double-velocity correlation, $C = 3/2$ (Kolmogorov, 1991a, p. 17).

Kolmogorov did not derive the spectral law for the distribution of energy $E(k) \sim k^{-5/3}$ (k is the wave number) which is commonly referred to by his name—this law was published in the same year by his pupil Obukhov in a paper "On the spectral energy distribution in a turbulent flow" (Obukhov, 1941). Therefore it is more appropriate to call it "Obukhov's 5/3-law".[12] In the same vein it has been suggested to replace the label "K41" by "KO41" in order to honor Obukhov's contributions (Falkovich, 2011, pp. 215–220). "Kolmogorov himself emphasized often that only the statement of the problem is due to him," Yaglom recalled many years later, "while the main quantitative results were obtained independently by him and Obukhov dur-

[11] The "four-fifths law" was later regarded as one of the most important results (Frisch, 1995, Sect. 6.2).

[12] Mathematically it is equivalent with the two-thirds law, see Frisch (1995, Sect. 6.3.1).

ing a period when they were separated. Therefore it is not surprising that the approach used by the two authors was quite different" (Yaglom, 1990, pp. vi–vii).

In January 1942 Kolmogorov presented a report on his results at a meeting in Kazan, a town on the Volga where the Academy had been evacuated in the war. In the discussion Lev Davidovich Landau (1908–1968) praised Kolmogorov as "the first to provide correct understanding of the local structure of a turbulent flow" but he expressed some doubts about the equations of turbulent motion that Kolmogorov presented at this occasion (Spalding, 1991, p. 216). Two years later, he was even more critical. In a note "On the problem of turbulence" Landau remarked that despite extensive studies on turbulent motions "the very essence of this phenomenon is still lacking sufficient clearness" (Landau, 1944, p. 387).

Outside Russia Kolmogorov's theory became known only after the Second World War. George Batchelor (1920–2000), a student of Taylor, scrutinized the English-language edition of the Proceedings of the USSR Academy of Science in the beginning of his own investigation on turbulence. "Like a prospector systematically going through a load of crushed rock, I suddenly came across two short articles, each about four pages in length, whose quality was immediately clear," Batchelor recalled many years later his first encounter with Kolmogorov's theory (quoted in Moffatt, 2002, p. 30). When Batchelor asked Taylor whether Kolmogorov's theory was known to him before 1941, "implicitely if not in the form in which he put it forward", Taylor responded:

> I did realize as far back as 1917 that there must be something that gives small-scale turbulence a statistically isotropic character and that this would be a result of some universal quality in the grinding-down process. I could see that for separations of pairs of points outside the range in which velocity gradients can be regarded as constant there should be a range in which a universal law of grinding down determines the relationship between the correlations and the separation of two points in a turbulent field. However I did not see how to turn this idea into a mathematical description which could form the basis of a theory and could predict things that could be verified or disproved experimentally. [...]
>
> Though in 1937 I had realized the equivalence of the correlation description of turbulence and the spectrum description, my idea of the dynamics was directed to trying to connect the rate of increase of mean-square vorticity with dispersion, because if two neighbouring points on a vortex line are separating the vorticity is increasing, and of course the rate of dissipation of energy is increasing. This idea was expressed in a suitable mathematical form by Kármán and Howarth for isotropic turbulence, but it did not lead to the law of grinding down in scale of eddies and I do not think I had any idea that a similarity argument could be used for this purpose till 1945 when you discovered those 1941 papers by Kolmogoroff in a library. (Quoted from Batchelor, 1975, pp. 630–631)

Taylor's concerns about the vorticity and the "grinding down" of eddies were not laid to rest by Kolmogorov's theory. Landau's remark that "the very essence of turbulence" remained unclear also called for further investigations.[13] Batchelor devoted a good deal of his own career to the dissemination and further elaboration of Kolmogorov's theory (Moffatt, 2011, Sect. 8.4). Not least due to Batchelor's

---

[13] "Landau's objection" became subject of expert debate still many years later, see Frisch (1995, Sect. 6.4) and Moffatt (2011, p. 278).

missionary zeal the statistical theory of turbulence reached center-stage after the Second World War (see Sect. 11.1).

Chapter 8
The Onset of Turbulence

Abstract By the early 1920s, the failure of the Orr–Sommerfeld approach to yield a critical limit at which a laminar flow becomes turbulent gave rise to controversial debates. Experimental evidence for the role of instabilities was lacking. In view of more tangible causes for the onset of turbulence in the boundary layer, like the fluctuations of the air stream in a wind tunnel, the transition to turbulence as postulated by the Orr–Sommerfeld approach was contested. In the early 1940s experiments in wind tunnels with low turbulence at the National Bureau of Standards and the Guggenheim Aeronautical Laboratory of the California Institute of Technology (motivated by the quest for laminar airfoils with reduced drag) finally corroborated the theoretically predicted boundary layer instability. A revision of the stability theory by Kármán's doctoral student Lin rehabilitated the contested Orr–Sommerfeld approach as a viable route for analysing the transition to turbulence.

8.1 Heisenberg's Doctoral Work

When Landau considered the turbulence problem not solved by Kolmogorov's theory he suggested that "the problem may appear in a new light if the process of initiation of turbulence is examined thoroughly" (Landau, 1944, p. 387). Thus he re-oriented the focus on "the turbulence problem" as it was perceived already long ago, i.e. from the vantage point of the onset of turbulence. Landau's own approach and its ramifications are subject of the post-war history of turbulence (see Sect. 13.4)—here it reminds us to resume the history of the stability problem that we have left as it was discussed in the early 1920s (see Sect. 5.5).

All attempts to derive a critical Reynolds number at which a laminar flow becomes turbulent had failed—this was the sober resumée in Noether's review of "The turbulence problem" (Noether, 1921). The analysis had been confined to plane flows that could be regarded as approximations of laminar boundary layer flow along a flat plate or the flow between two plates moving in opposite directions (plane Couette flow). But neither the ideal flow approximation à la Rayleigh nor the more general Orr–Sommerfeld approach yielded the desired instability at a critical Reynolds number. The flow profiles which Prandtl and Tietjens analysed by Rayleigh's approach

M. Eckert, *Turbulence—an Odyssey*, History of Physics,
https://doi.org/10.1007/978-3-030-91459-2_8

yielded instability when stability was expected, whereas the analysis of plane Couette flow by Sommerfeld and his disciple Hopf yielded stability for all Reynolds numbers and disturbances (Eckert, 2010).

Despite these failures (or because of them) the problem remained a paramount challenge for aspiring theorists. In May 1922 Hopf informed Prandtl that "we may be in for a sensation" because one of Sommerfeld's "prodigies" seemed to have mastered the problem of turbulence.[1] The prodigy was Werner Heisenberg (1901–1976), then a student in his fifth semester. After showing his exceptional talents in the field of atomic theory (and before his discoveries in quantum mechanics) he chose the turbulence problem as a subject of his doctoral thesis. "The hydrodynamic explanation of turbulent flow in tubes or channels is a problem that is famous for its difficulty," Sommerfeld introduced his report about Heisenberg's dissertation, "I could not have proposed such a difficult theme to another one of my disciples."[2]

Sommerfeld spent the time when Heisenberg elaborated his dissertation as guest-professor in the USA. He had sent his doctoral students for this period to Göttingen University, where they could discuss atomic theory with Max Born (1882–1970) and turbulence with Prandtl. Heisenberg kept Sommerfeld informed about his progress by correspondence. "I must tell you in more detail about the turbulence because it now begins to become more reasonable," he wrote in October 1922 to Sommerfeld.[3] By January 1923 he reported exciting news about recent investigations in atomic theory, but no news about his work on turbulence that he had recently resumed after some pause. "However, I haven't got any further, the mathematical difficulties are, for me at least, very great."[4]

The concept of Heisenberg's thesis was based on the Orr–Sommerfeld approach, but unlike his predecessors he resorted to other series expansions with regard to the Reynolds number and the wave number of the disturbance superposed to the laminar flow. "In the free handling of these approximations lies the strength and weakness of the work," Sommerfeld discerned the main challenge for Heisenberg's analysis. "It would have been impossible to achieve the many remarkable results if the author would have wished to stick to precise error estimates." These concerned not only the calculation of the limit of stability for plane Poiseuille flow, but also fully developed turbulence. The latter could be addressed because Heisenberg's approach dealt with

---

[1] "Übrigens steht uns vielleicht eine Sensation bevor: Sommerfeld schrieb, dass eines seiner Wunderkinder (Heisenberg, im 5. Semester) die absoluten Maße der Kármánschen Wirbelstraßen berechnet und angeblich auch das Problem der Turbulenz gemeistert habe." Hopf to Prandtl, 26 May 1922. AMPG, Abt. III, Rep. 61, Nr. 704.

[2] "Die hydrodynamische Erklärung der Turbulenz von Strömungen in Röhren oder Kanälen bildet ein Problem, das wegen seiner Schwierigkeit berühmt ist. [...] Ich hätte keinem anderen meiner Schüler ein Thema von dieser Schwierigkeit als Dissertation vorschlagen können." Sommerfeld, report to the faculty, 23 July 1923. UAM, OC-I-49p.

[3] "Über die Turbulenz muß ich Ihnen etwas ausführlicher erzählen, weil sie wirklich jetzt allmählich sehr vernünftig wird." Heisenberg to Sommerfeld, 17 October 1922. DMA, HS 1977-28/A, 136.

[4] "Hoffentlich geht es mit der Turbulenz, ich hab in der letzten Zeit wieder daran gearbeitet; weiter bin ich aber noch nicht gekommen, die mathematischen Schwierigkeiten sind, für mich wenigstens, sehr groß." Heisenberg to Sommerfeld, 15 January 1923. DMA, HS 1977-28/A, 136.

laminar and turbulent flow on the same footing, both subject to the Navier–Stokes equations. "In the handling of the present problem, Heisenberg shows once again his extraordinary abilities: complete command of the mathematical apparatus and daring physical insight."[5]

But Heisenberg's oral doctoral examination was a fiasco (Rechenberg, 2010, Sect. 2.4). It almost ruined his career. Heisenberg displayed an embarrassing ignorance about subjects that students learn in lecture courses and practica on experimental physics—to which Heisenberg had paid little attention in the course of his unusual study. The professor of experimental physics, Wien, was annoyed and wanted to flunk Heisenberg. He only passed the exam thanks to Sommerfeld's intercession (Cassidy, 2009, pp. 117–119). Heisenberg's thesis, too, did not meet with unanimous praise. Prandtl felt that "the last word in this matter is not yet spoken" and dismissed him with lukewarm encouragment. "However, your work will undoubtedly, even if parts of it are later refuted, represent a very good step forward on the way to understanding turbulence."[6] When Heisenberg prepared the publication of his thesis in the *Annalen der Physik* he confided to Sommerfeld that Prandtl still had some reservations because "the real phenomena are many times more complicated than my approaches." But Heisenberg was confident "that at least qualitatively the calculations are correct. I also have the impression from my discussions with Prandtl that a number of unrelated stability questions have been reconciled through my work and have found a reasonable answer."[7]

Heisenberg submitted his paper "On Stability and Turbulence of Fluid Flows" in January 1924 to the *Annalen der Physik*. The main result was the derivation of a limiting curve for the Reynolds number R as a function of the disturbance α which displays the limit of stability for plane Poiseuille flow (Fig. 8.1). However, Heisenberg's theory yielded only part of the stability curve. He summarized the result rather qualitatively by the following conclusions: First, there is a maximal α and a minimal R for the limit of instability. Second, for a definite R there is both a maximal and minimal α; within this range there is instability, outside stability. Third, the maximal value of α is about 0.7 ($\alpha^2 \approx 1/2$), the minimal value of R is of the

[5] "In der freien Handhabung der hierdurch gegebenen Annäherungen liegt die Stärke und Schwäche der Arbeit. Es wäre unmöglich gewesen, zu den vielen bemerkenswerten Resultaten zu gelangen, wenn der Verfasser sich an genaue Fehlerabschätzungen hätte halten wollen. [...] An der Behandlung des gegenwärtigen Problems zeigt er seine ausserordentlichen Fähigkeiten von Neuem: volle Beherrschung des mathematischen Apparates und kühne physikalische Anschauung." Sommerfeld, report to the faculty, 23 July 1923. UAM, OC-I-49p.

[6] "... dass nach meiner Ansicht das letzte Wort in der Sache noch nicht gesprochen ist. Ihre Arbeit wird aber zweifellos, selbst wenn Teile davon später widerlegt werden sollten, einen sehr guten Schritt vorwärts auf dem Wege zur Erkenntnis der Turbulenz bedeuten." Prandtl to Heisenberg, 19 July 1923. AMPG, Abt. III, Rep. 61, Nr. 643.

[7] "Allerdings betont Prandtl mit Recht auch immer wieder, daß die wirklichen Erscheinungen noch um ein vielfaches komplizierter sind, als meine Ansätze. Nach Photografien ist das auch gar nicht zu bestreiten, aber ich glaube doch, daß wenigstens qualitativ die Rechnungen das richtige liefern. Auch hab ich aus den Gesprächen mit Prandtl den Eindruck, daß durch meine Arbeit doch eine Reihe von unzusammenhängenden Stabilitätsfragen unter einen Hut gebracht sind und eine vernünftige Antwort finden." Heisenberg to Sommerfeld, 8 December 1923. DMA, HS 1977-28/A, 136.

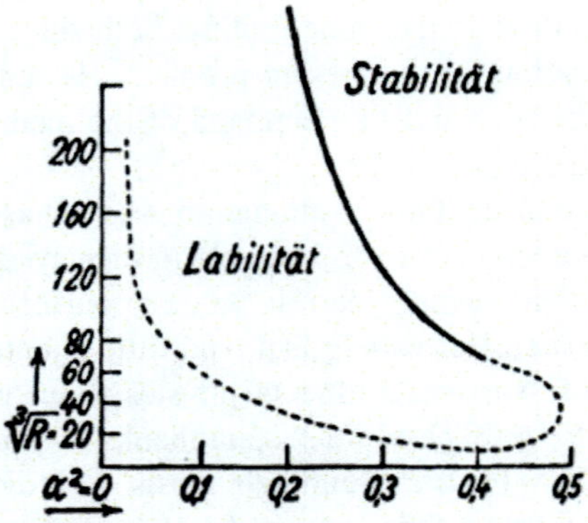

Fig. 8.1 Heisenberg's limiting curve for the stability of plane Poiseuille flow. α is the wave number of the disturbance, R the Reynolds number. "The minimal value of R is in the order of magnitude 10^3," he estimated the critical limit. "A sufficiently precise calculation of this minimal value from the figure is not possible" (Heisenberg, 1924, pp. 604–605)

order 10^3; "a sufficiently precise calculation of this minimal value from the figure is not possible" (Heisenberg, 1924, pp. 604–605).

In the second part of his dissertation Heisenberg conceived an approach for developed turbulence. He accounted for the plane flow between two parallel walls in terms of a Fourier series of the stream function where the lower terms represented laminar flow and the higher terms turbulence. He aimed at the 1/7 law for the distribution of flow velocities close to the wall—but his analysis was ridden with more or less convincing series expansions. "That I have reservations about your derivation of the 1/7-law is known to you from oral discussions," Prandtl reacted to this part of Heisenberg's theory. He believed that this law could only be derived "on a statistical basis".[8]

Heisenberg's result on the stability of plane Poiseuille flow, however, indicated a breakthrough of the deadlock that had doomed the Orr–Sommerfeld approach for so long. But it was not regarded as a turning point. In June 1926, Noether published in *ZAMM* a proof that the Orr–Sommerfeld approach necessarily yields stability for all Reynolds numbers for the cases that had been analysed so far—which included Heisenberg's parabolic flow profile. Heisenberg's result was therefore "in contradiction" with Noether's analysis (Noether, 1926, p. 242). "Heisenberg's way of reasoning", Hopf wrote to Prandtl, "is lacking any mathematical foundation."[9] He arrived at the same verdict in a chapter on "Viscous fluids" in the *Handbuch der Physik* where he reviewed the "stability problem" in some detail (Hopf, 1927, pp. 128–134). But Noether's analysis also provoked controversial discussions. Prandtl

[8] "Dass ich gegen Ihre Herleitung des 1/7-Gesetzes Bedenken trage, ist Ihnen von mündlichen Besprechungen her bekannt. [...] Auch glaube ich, dass das 1/7-Gesetz auf statistischer Grundlage zustande kommt ..." Prandtl to Heisenberg, 19 July 1923. AMPG, Abt. III, Rep. 61, Nr. 643.

[9] "... Heisenbergs Schlussweise, der jedes mathematische Fundament fehlt." Hopf to Prandtl, 2 August 1926. AMPG, Abt. III, Rep. 61, Nr. 704.

admitted in his first reaction to Noether's paper that "my mathematical knowledge is not remotely sufficient to fully understand your calculations," but he expressed doubts about Noether's approximations, in particular with regard to the "parabolic profile".[10] Prandtl considered the stability of the parabolic profile probable according to pipe flow experiments, but "certainly not yet proven by Noether's calculations."[11]

8.2 The Tollmien–Schlichting Instability

Prandtl made his concerns about Noether's analysis also public in a short note "On the turbulence problem" in *ZAMM* (Prandtl, 1926b). If Noether was right, further efforts to derive a critical limit of stability of boundary layer flow by the Orr–Sommerfeld approach were doomed to failure. After Tietjens's analysis of straight-line profiles did not yield the desired result (see Sect. 5.5), Prandtl entrusted another doctoral student with a new stability analysis of boundary layer flow. The intention was "to study the oscillations of a profile composed by a parabola and a straight line", i.e. a velocity profile which was close to the profile of laminar boundary layer flow along a flat plate ("Blasius flow"). "This work I have had in mind for a long time, but it got stuck because a doctoral student failed," Prandtl informed Hopf about his plans in this regard. "However, I hope that we will be able to tackle it again soon."[12]

The failure of Prandtl's doctoral student in 1926 may have prevented another attempt by another doctoral student. Prandtl rather resorted to his experienced collaborator Tollmien at the Kaiser-Wilhelm-Institute. Tollmien had accomplished his doctoral work already in 1924 and just displayed his mathematical virtuosity anew by elaborating Prandtl's mixing length concept for a theory of the spreading of turbulent jets and wakes (see Sect. 6.2). Furthermore, Prandtl must have expected another rivalry with Kármán's group at Aachen when Hopf informed him about "a theoretical assistent who wants to do turbulence calculations".[13] The assistant was Theodor Sexl (1899–1967), a theoretical physicist from the University of Vienna, who would later direct his research towards nuclear physics but in the beginning of his career was intrigued by the hydrodynamic stability problem which he hoped to solve during

---

[10] "... denn meine mathematischen Kenntnisse reichen nicht entfernt hin, um Ihre Rechnungen ganz zu verstehen. [...] in welcher Weise Sie beim Parabelprofil mit der Vernachlässigung von α^2 vorgegangen sind." Prandtl to Noether, 12 July 1926. AMPG, Abt. III, Rep. 61, Nr. 1155.

[11] "... dass die Stabilität des Parabelprofiles, die ja wegen der Experimente wahrscheinlich ist, sicher durch die Noetherschen Rechnungen noch nicht erwiesen ist." Prandtl to Hopf, 20 July 1926. AMPG, Abt. III, Rep. 61, Nr. 704.

[12] "... dass wir die Absicht haben, die Schwingungen eines aus einer Parabel und einem geraden Stück zusammengesetzten Profiles zu studieren [...] Diese Arbeit habe ich schon lange vor, aber sie ist dadurch stecken geblieben, dass ein Doktorand versagt hat. Ich hoffe aber, dass wir sie demnächst wieder in Angriff nehmen können." Prandtl to Hopf, 20 July 1926. AMPG, Abt. III, Rep. 61, Nr. 704.

[13] "Ich bekomme jetzt – in Vertretung Kármáns – wahrscheinlich einen theoretischen Assistenten, der Turbulenzrechnungen machen möchte". Hopf to Prandtl, 2 August 1926. AMPG, Abt. III, Rep. 61, Nr. 704.

a research sojourn in Aachen. He published the results of his Aachen investigation in a sequence of three papers in the *Annalen der Physik*, beginning with an analysis "On the question of stability of Poiseuille and Couette flow" (Sexl, 1927a) and "On three-dimensional disturbances of Poiseuille flow" (Sexl, 1927b). By and large, Sexl's study seemed to confirm Noether's proof that the Orr–Sommerfeld approach would not yield a critical limit of stability. In the third paper he derived a special oscillatory solution for Poiseuille flow, but no transition to turbulence. Heisenberg's effort to derive a limit for the onset of turbulence appeared once more as a failure due to the insufficient mathematical foundation (Sexl, 1928, p. 571).

Sexl's study of Couette and Poiseuille flow did not directly concern boundary layer flow, but Tollmien must have expected to meet with the same verdict because his analysis was also based on the Orr–Sommerfeld approach. He presented the result of his painstaking analysis only two years later in March 1929—and by setting himself apart from Heisenberg who did not resolve the question of convergence of his approximation and contented himself with conjectures about the limits of stability. Unlike Heisenberg, Tollmien could present the entire limiting curve that separated stable and unstable flows (Tollmien, 1929, p. 43) (Fig. 8.2).

The lowest critical limit was found at a Reynolds number of 420—a value that roughly agreed with the transition from laminar to turbulent boundary layer flow as measured in Burgers's wind tunnel at Delft (Burgers, 1925) and more recently in Kármán's laboratory at the TH Aachen (Hansen, 1928). But it was not clear whether the transition to turbulence (determined in the experiments by the change of the velocity gradient of the boundary layer) agreed with the limit of stability. Furthermore, the curve that separated stable from unstable states could not be measured without knowledge of the wave number of the disturbance. Direct corroboration of Tollmien's theory by experiments was therefore insufficient, at best, if not impossible (Tollmien, 1929, p. 43).

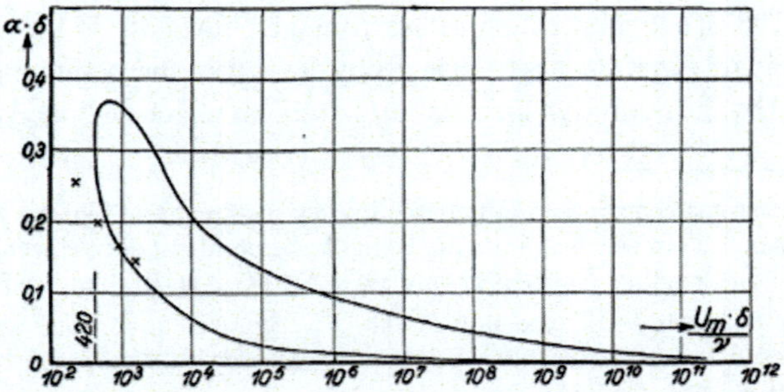

Fig. 8.2 Tollmien's limiting curve for the stability of boundary layer flow along a flat smooth plate. The exact "Blasius flow" profile was approximated by a linear piece near the wall that connected to a parabola at some distance from the wall. δ is a measure for the thickness of the boundary layer, α is the wave number of the disturbance, $U_m\delta/\nu$ is the Reynolds number related to δ. The unstable states of flow lie within the curve (Tollmien, 1929, p. 41)

Tollmien's analysis benefited from the results of Tietjens's paradoxical finding that flows which were stable according to Rayleigh's criterion of frictionless flows became unstable by adding viscosity (Tietjens, 1925). Tollmien clarified this paradox by showing that the influence of viscosity must not only be taken into account in the immediate vicinity of the wall, but also in the vicinity of a critical layer, where the wave propagation speed of the disturbance coincides with the speed of the main flow. Tollmien's result suggested a new analysis of other cases that had been earlier classified as stable. If plane Couette flow was analysed during the start-up phase the stability investigation had to be based on a profile which occurs when one wall is set in motion from rest to a constant velocity. The linear velocity profile represents only the final steady state of plane Couette flow. Prandtl therefore asked Schlichting, another collaborator at the Kaiser-Wilhelm-Institute, to perform the stability analysis of plane Couette flow along the lines of Tollmien's study with a transient velocity profile from the start-up period. Schlichting obtained for such transient velocity profiles similar stability curves as Tollmien for the boundary layer profile, and in the limit a critical Reynolds number (with respect to the distance between the walls) of 19,300. This is how he summarized his analysis:

> In this paper, the stability investigation is not based on the stationary, developed profile with linear velocity distribution, but the turbulence generation is treated as a start-up effect. One arrives at a critical Reynolds number for the Couette flow if one examines those laminar profiles which arise during the temporal development of the boundary layer when one wall is very rapidly displaced from rest into constant velocity in a tangential direction. [...] The determination of the critical Reynolds number amounts to the solution of an eigenvalue problem of the fourth order perturbation differential equation. Only a relatively small range of wavelengths of the superimposed perturbations proves to be dangerous for the laminar flow.[14]

After this study Schlichting focused on the boundary layer flow along a flat smooth plate. Tollmien had only determined the curve that separated stable from unstable states. Schlichting now attempted to compute the growth of the unstable disturbances at various Reynolds numbers with respect to their wavelength in order to account more closely for the onset of turbulence in plate flow. "An explicit calculation of the magnitude of the onset for every given state of flow and every wavelength of disturbance has not yet succeeded," Schlichting cautioned, "but it is possible by further developing the earlier methods of calculation to make statements about the growth of those disturbances that lie near the stability limit."[15] Schlichting expressed

---

[14] "Es wird in dieser Arbeit der Stabilitätsuntersuchung nicht das stationäre, ausgebildete Profil mit linearer Geschwindigkeitsverteilung zugrunde gelegt, sondern die Turbulenzentstehung wird als ein Anlaufeffekt aufgefasst. Man gelangt zu einer kritischen Reynoldsschen Zahl für die Couetteströmung, wenn man diejenigen Laminarprofile untersucht, die bei der zeitlichen Entwicklung der Grenzschicht entstehen, wenn die eine Wand aus der Ruhe heraus sehr plötzlich in konstante Geschwindigkeit in tangentieller Richtung versetzt wird [...] Die Bestimmung der kritischen Reynoldsschen Zahl kommt auf die Lösung eines Eigenwertproblems der Störungsdifferentialgleichung 4. Ordnung hinaus. Nur ein verhältnismässig kleiner Wellenlängenbereich der überlagerten Störungen erweist sich für die Laminarströmung als gefährlich." (Schlichting, 1932, pp. 934–935).

[15] "Eine explizite Berechnung der Größe der Anfachung für jeden vorgegebenen Strömungszustand und jede vorgegebene Störungswellenlänge ist uns zwar bisher noch nicht gelungen, jedoch lassen

the hope that his results are helpful for experimental investigations of the onset of turbulence.

Two years later Tollmien resumed the theoretical debate of stability investigations and elaborated a general criterion about the instability of plane laminar flows (Tollmien, 1935). It extended Rayleigh's classical "turning point" theorem for the stability of ideal flows. While Rayleigh had proven only the existence of a turning point in the flow profile as a necessary condition for the occurrence of unstable oscillations, Tollmien showed that the existence of a turning point is also a sufficient condition for the existence of such oscillations.

8.3 Disputes on Hydrodynamical Stability

Yet the work of Tollmien and Schlichting did not meet with unanimous approval. In a review on "Hydrodynamical Stability" at the Semicentennial of the American Mathematical Society, their contributions were completely ignored. The reviewer was John L. Synge (1897–1995), head of the Department of Applied Mathematics at the University of Toronto. Even Heisenberg's and Noether's studies were only mentioned as examples of mutually contradictory theories. The attempt to account for the onset of turbulence in terms of stability theory appeared almost futile: "It represents mathematical problems of no small difficulty: triumphs are few and disappointments many" (Synge, 1938, p. 227). Both plane Couette flow and plane Poiseuille flow appeared stable according to the Orr–Sommerfeld approach. Even "Poiseuille motion in a tube of circular section is stable for axially symmetric disturbances of any wavelength for any value of the Reynolds number," the reviewer summarized Sexl's application of the Orr–Sommerfeld approach to three-dimensional Poiseuille flow in a pipe (Synge, 1938, p. 255). The conclusion seemed clear that the onset of turbulence could not be explained by the Orr–Sommerfeld approach.

What about other approaches on the stability of flows? Taylor had published in 1923 a thorough analysis of circular Couette flow, i.e. the rotation of a fluid confined between two rotating coaxial cylinders instead of two parallel plane walls. Taylor assumed disturbances with rotational symmetry (unlike the travelling plane waves moving with the main flow in the Orr–Sommerfeld approach). He had derived a critical limit beyond which toroidal vortices (later called Taylor vortices) develop (Taylor, 1923). The agreement of theory and experiment was remarkable. But Taylor's result only concerned the transition between different stable states of flow. "A clarification of the actual significance of Taylor's investigations for the turbulence

---

sich durch Weiterentwicklung der früheren Rechenmethoden Aussagen machen über die Anfachung derjenigen Störungen, die in der Nähe der Stabilitätsgrenze liegen." (Schlichting, 1933, pp. 182–183).

problem, however, has not yet taken place", Sexl remarked in his stability analysis of Poiseuille and Couette flow.[16]

Was the alleged failure of the Orr–Sommerfeld approach due to the neglect of three-dimensional disturbances? This question was settled in 1933 by Herbert Squire (1909–1961), a British mathematician and aeronautical engineer. Squire had just spent a research sojourn with Prandtl in Göttingen where he became familiar with the long history of the Orr–Sommerfeld approach, in particular the contradicting results of Noether and Tollmien. Squire proved for flows between parallel walls that independent of the specific velocity profile "it is sufficient to confine attention to disturbances of two-dimensional type." He showed that "if any velocity profile is unstable for a particular value of Reynolds' number, it will be unstable at a lower value of Reynolds' number for two-dimensional disturbances" (Squire, 1933, pp. 627–628). In other words: two-dimensional disturbances were more harmful for the stability of a flow than three-dimensional ones. Thus the latter could not resolve the disputed issue whether a flow was stable or not.

While hydrodynamical stability theory was at pains to explain the onset of turbulence in terms of the Orr–Sommerfeld approach, experiments contributed little evidence for the role of instabilities. "A clear decision for or against Tollmien's theory is still pending," Nikuradse confessed after a series of experiments in a water channel in Prandtl's Kaiser-Wilhelm-Institute.[17] Pipe flow measurements from Schiller's institute in Leipzig (see Sect. 5.5) nourished the view that the onset of turbulence is essentially caused by vortices from the inlet rather than instabilities inside the pipe. In 1934 Schiller published the results of two doctoral theses from his Leipzig institute on pipe and channel flow, respectively, which confirmed this view with quantitative data. "The critical wavelengths that we observe," Schiller remarked in closing, "are not quite consistent with the Tollmien–Schlichting theory."[18] Hot-wire measurements of the boundary layer along a flat plate in Dryden's laboratory at the National Bureau of Standards suggested that "the fluctuations in the laminar layer are to be regarded as forced oscillations produced by the turbulence of the wind-tunnel air stream" (Dryden, 1936, p. 356).

In September 1938 Taylor reviewed the contemporary research front on this issue at the Fifth International Congress of Applied Mechanics in Cambridge, Massachusetts. "The only case where instability has been proved mathematically, and also observed, is that of ring shaped disturbances between rotating cylinders," he distinguished the well established instability of this flow from the unproven Tollmien–Schlichting instability. Although he did not completely rule out the existence of "Tollmien's waves", he regarded them as an unlikely mechanism for the transition

[16] "Eine Aufklärung der eigentlichen Bedeutung der Taylorschen Untersuchungen für das Turbulenzproblem ist allerdings bisher nicht erfolgt." (Sexl, 1927a, p. 848).

[17] "Eine klare Entscheidung für oder gegen die Tollmiensche Theorie steht hiernach noch aus." (Nikuradse, 1933, p. 176).

[18] "Die von uns beobachteten kritischen Wellenlängen stehen [...] nicht recht mit der Tollmien–Schlichtingschen Theorie in Einklang." (Schiller, 1934).

to turbulence. So far "the experimental evidence is against the instability theory". Instead he discerned the following four causes:

> In summing up the present situation it seems to me that transition may be determined by conditions at the leading edge, by the state of the surface, by the turbulence in the stream when there is a reverse pressure gradient, or by conditions arising in the boundary layer after separation. These effects are in general superposed so that it is difficult to fix on one of them as the sole cause of transition. It does not seem to me that the mechanics of any of these four causes is properly understood, but recent improvements in our technical equipment, wind tunnels with very low turbulence for instance, are now enabling us to arrange our experiments so that they can be separated. (Taylor, 1939, p. 308)

8.4 A Kind of Boundary Layer "Flutter"

In a low-turbulence wind tunnel the boundary layer along an edgewise exposed plate would not become turbulent by eddies from the air stream so that an external cause for the transition to turbulence could be ruled out. If the eddying air stream was the major cause for the onset of turbulence the boundary layer would remain laminar for a longer distance from the leading edge of the plate—or an appropriately curved airfoil. The skin friction of such "laminar airfoils" was reduced to the extent that the point of transition to turbulence was moved downstream from the leading edge. Reduction of wind tunnel turbulence, therefore, was key for the development of airfoils with a minimal drag due to the lower skin friction of laminar flow. Measurements in a new low-turbulence tunnel of the NACA showed that the drag of appropriately shaped laminar airfoils could be considerably reduced. On the eve of World War II, this result came in time for the design of fighter airplanes. "Discovery during the past year of a new principle in airplane-wing design may prove of great importance," the NACA revealed in its Annual Report for 1939. "The transition from laminar to turbulent flow over a wing was so delayed as to reduce the profile drag, or basic air resistance, by approximately two-thirds." (Quoted in Hansen, 1987, p. 116).

The design of low-turbulence wind tunnels was therefore primarily motivated by the search for laminar airfoils with a minimal drag and less by investigations of boundary layer stability. Dryden's Aerodynamics Division at the NBS pursued a number of projects under contract for the NACA in this quest. In May 1940, Dryden's research agenda comprised "Transition Phenomena at Low Turbulence", "Construction of Turbulence Measuring Equipment for Langley Field" and "Methods of Reducing Wind Tunnel Turbulence". The purpose of the first item on this list was to determine "the effect of small intensities of wind tunnel turbulence on the position of transition from laminar to turbulent boundary-layer flow". For these measurements a flat aluminium plate was exposed to the air stream of the low-turbulence tunnel at the NBS. The transition to turbulence was determined "by surface pitot [i.e. pressure measurements] and hot wire". Since other measurements had shown that a transition can be caused even at very low levels of wind tunnel turbulence, the focus was placed "on reducing the turbulence as much as possible and investigating transition at the lowest values obtainable".[19]

Neither the NACA nor Dryden and his collaborators at the NBS aimed originally at an experimental test of the stability theory. The observation of Tollmien–Schlichting waves was reported in May 1941 as an unexpected side-effect:

> Under conditions of low stream turbulence, transition is found to result from growth in amplitude of a fairly regular oscillation of the boundary layer. The oscillation was in fact so regular that considerable work was done to eliminate the possibility that it might be due to mechanical vibration. The frequency was dependent on the speed of the stream and the thickness of the boundary layer. The oscillations have some of the characteristics of those predicted by the Tollmien–Schlichting theory which have not heretofore been experimentally observed. The frequencies are in fact in line with those predicted. Likewise the speed fluctuation in the outer portion of the boundary layer is in opposite phase to that near the wall as predicted by the theory. [...] The natural oscillations of the boundary layer, a kind of boundary layer "flutter", appear to be the key to the explanation of transition and further study may be expected to unravel the various factors which control the appearance, growth, and damping of these oscillations.[20]

Galen B. Schubauer (1904–1992) and Harold K. Skramstad (1908–2000), the experimenters in Dryden's group who conducted these investigations, were rather surprised by the regular sinusoidal hot-wire recordings which preceded the erratic traces after the onset of turbulence (see Fig. 8.3). The designation of the phenomenon as "flutter" alluded to wing flutter, a notorious aeronautical problem that was investigated in wind tunnel tests. For the next fiscal year the common NBS-NACA research on transition phenomena at low turbulence was titled "Investigation of Laminar Boundary-Layer Oscillation." Only now the "natural oscillations of the boundary layer which have been studied by Tollmien and Schlichting" received full attention as an investigation in its own right. NACA commissioned basic boundary-layer research at the height of the war only on condition that the results would be "applicable to current problems relative to military aircraft both to wing development and to ducting problems as well," the headquarters in Washington D.C. advised the Langley Laboratory in February 1943 (Roland, 1985, vol. 2, p. 549).

The NACA also sponsored investigations on the laminar–turbulent transition in boundary layers at the GALCIT in Pasadena, where von Kármán directed an ambitious group of theorists and experimenters. Here, too, the emphasis was on aeronautical applications. In 1938, a small wind tunnel was built and adjusted to low-turbulence studies in a similar manner as Dryden's low-turbulence wind tunnel at the NBS. Kármán entrusted a student, Hans Liepmann (1914–2009), with measurements on the stability of the laminar boundary layer on curved surfaces in the GALCIT low turbulence tunnel. Many years later, Liepmann recalled the moment when Kármán learned about the results obtained by Dryden's group:

> Sometime in 1941 with the war in Europe in its second year, I came up the narrow substandard stairs in the Guggenheim laboratory. At the top of the stairs I met Clark Millikan [1903–

[19] Lewis to Warner, 6 May 1940. Dryden papers, Special Collections, Milton S. Eisenhower Library, The Johns Hopkins University, Misc. Correspondence, re. NACA.

[20] NBS to NACA, 10 May 1941. Dryden papers. Misc. Corresp. 1941 re. NACA.

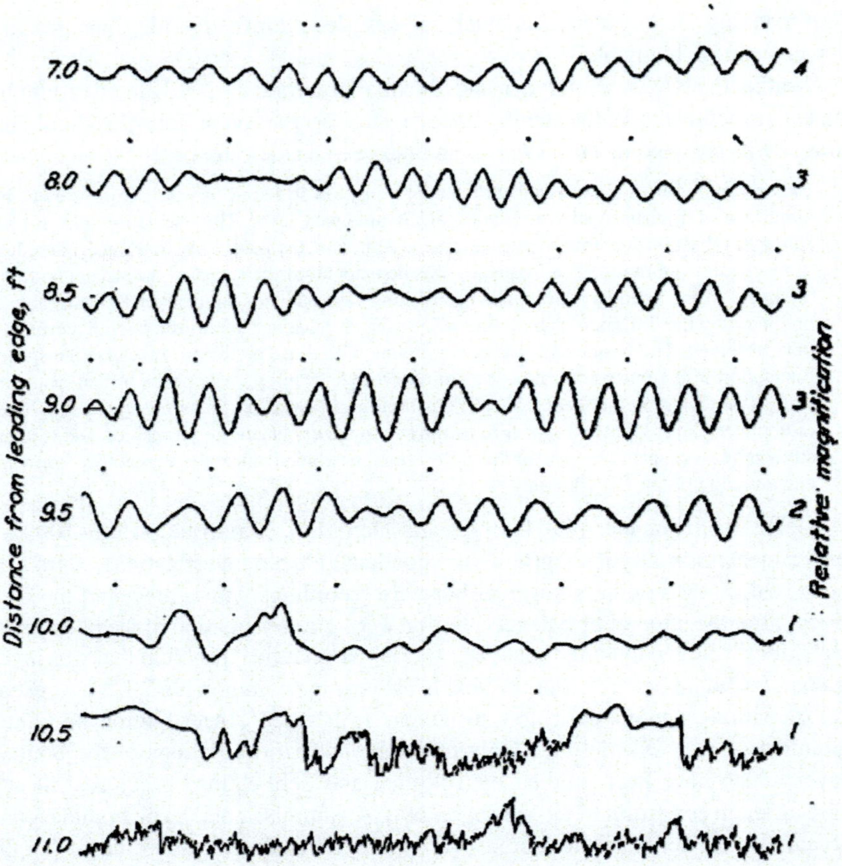

Fig. 8.3 Oscillogramm of boundary layer oscillations measured by hot-wire probes along a flat plate in the wind tunnel at the National Bureau of Standards (Schubauer and Skramstad, 1943, p. 334)

> 1966] handing a sheaf of papers to von Karman with the words "It's a complete German victory!" I was stunned. Fortunately, however, the victory was not another one won by Hitler but referred to the experimental verification of the laminar instability theory of Heisenberg and Tollmien by Schubauer and Skramstad at the National Bureau of Standards (NBS). I still vividly remember the impact these experiments had on me; they forcefully demonstrated to me the beginning of a new area in transition research. I immediately set out to reproduce the NBS results. (Liepmann, 1997, p. R1)

A year later, Liepmann also found the oscillations predicted by the Tollmien–Schlichting theory. Both the NBS and GALCIT results were summarized in 1943 in Advance Confidential Reports for the NACA. "The work is thus experimental confirmation of a mathematical theory of stability which had been in the process of development for a period of approximately 40 years, mainly by German investigators", the NBS-experimenters summarized their result (Schubauer and Skramstad,

1943, p. 327). "Measurements of the characteristics of the laminar oscillations agree closely with the results of Schubauer and Skramstad", Liepmann confirmed the discovery (Liepmann, 1943, p. 1).

Thus the disputed Tollmien–Schlichting instability was corroborated twice in the course of wartime research in American laboratories. A Chinese student of Kármán, Chia-Chiao Lin (1916–2013), added further evidence in the form of a theoretical investigation of the stability theory which became the subject of his Ph.D. dissertation. Lin's revision rehabilitated the disputed Orr–Sommerfeld approach. Lin was at that time not yet a naturalized US citizen and thus not cleared for confidential war research. Therefore, unlike Liepmann's experimental investigation, his theoretical analysis was not performed under a NACA war research contract. However, Lin knew about the experimental results and thanked Liepmann in the acknowledgement for valuable suggestions. Kármán sent Dryden a diagram "showing Lin's curve (present calculation) for the stability limit in the boundary layer compared to Schlichting and the experimental evidence." Lin's diagram confirmed the interpretation of the experiments in Kármán's and Dryden's laboratories. "Unfortunately because of the confidential character of yours and Liepmann's work this diagram cannot be published", Kármán regretted the wartime secrecy. "However I thought you might be interested in having a copy for your file."[21] The episode was also vivid in Liepmann's recollection: "CC Lin who, while at GALCIT, cleaned up and extended the mathematics of the theory had the advantage of knowing the experimental results which at the time of his research were still considered confidential and not published in the open literature" (Liepmann, 1997, p. R4).

The discovery of "Heisenberg–Tollmien or Tollmien–Schubauer waves", as Liepmann would have preferred to name this phenomenon (Liepmann, 1997, p. R4) instead of the now common designation "Tollmien–Schlichting waves", was disclosed only after the war. In retrospect it is just one of several instabilities which can play a role for the transition to turbulence. From a historical perspective the discovery of Tollmien–Schlichting waves stands out as the major impulse that rehabilitated the discredited Orr–Sommerfeld approach—and fuelled new investigations about the transition to turbulence after the war.

[21] Kármán to Dryden, 26 February 1944. Dryden papers, folder "1935–1959. Misc. Dryden–Von Kármán Correspondence".

Part III
Staging Turbulence

Turbulence research developed its profile gradually at a variety of places, institutions and conferences. This part follows the growth of the turbulence community through these stages up to a conference at Marseilles in 1961 that marked the beginning of turbulence as an international research field in its own right.

Chapter 9
The Rise of Applied Mechanics

Abstract Before the 1920s there was no common international stage for presenting research on turbulence. One group of researchers concerned with turbulence comprised "scientific engineers" trained in polytechnic schools, technical universities or similar institutions. They exerted their profession in disciplinary environments such as hydraulics, naval architecture or aeronautics. Theoretical studies on turbulence were also made by theoretical physicists and mathematicians. The results of these diverse scientific communities were presented without an explicit focus on turbulence in a heterogeneous variety of journals and at haphazard occasions. The first stages for presenting turbulence research in a more coherent manner were provided since the early 1920s under the label of applied mathematics and mechanics. A climax was reached in 1938 with a symposium on turbulence at the Fifth International Congress for Applied Mechanics in Cambridge, Massachusetts.

9.1 Turbulence Research Without a Common Label

Adhémar Jean Claude Barré de Saint-Venant and Joseph Boussinesq in France, Gotthilf Hagen and Hermann von Helmholtz in Germany, William Thomson (Lord Kelvin), John William Strutt (Lord Rayleigh) and Osborne Reynolds in Great Britain—to name only a few whose contributions are regarded as landmarks of nineteenth century research on turbulence—neither perceived nor presented their work under the label "turbulence". There was no common designation for turbulent flows before the 1920s. Saint-Venant and Boussinesq aimed for a "théorie des tourbillons liquides" and used expressions like "mouvements tumultueux" or "agitation tourbillonnaire". Even Reynolds never used the label "turbulence"; he described turbulent motions in terms like "sinuous paths" or "irregular eddies". In 1884 he compared "the two manners of motion of water" with the ordered and disordered movements of troops. The former was analogous to "direct or steady motion" in fluids, the latter corresponded to "sinuous or unsteady motion". Among the pioneers of turbulence research in the nineteenth century the term "turbulence" made its appearance presumably for the first time in 1887 in a paper by Kelvin on the "Stability of motion" of river flow (Schmitt, 2017).

M. Eckert, *Turbulence—an Odyssey*, History of Physics,
https://doi.org/10.1007/978-3-030-91459-2_9

The lack of a common label, of course, does not mean that turbulence was ignored as subject of closer investigation. It rather hints at the diversity of gateways to turbulence investigations (see part I). Before and to some extent well into the twentieth century, turbulent pipe and channel flow was the domain of hydraulic engineers; atmospheric flow was investigated by meteorologists; drag investigations were conducted by naval architects; hydrodynamics belonged to the realm of mathematicians and natural philosophers. Whenever the phenomenon of turbulence was encountered and investigated it was presented and discussed from the vantage point of one or another of these gateways. This is mirrored in the titles of pertinent monographs such as Boussinesq's *Essai sur la Théorie des Eaux Courantes*, Henry Darcy's *Recherches Expérimentales Relatives au Mouvement de l'Eau dans les Tuyaux*, Gustave Eiffel's *Recherches expérimentales sur la résistance de l'air exécutées à la tour Eiffel*, Horace Lamb's *Hydrodynamics*, Lorentz's *Abhandlungen über theoretische Physik* or von Mises' *Elemente der technischen Hydromechanik.*

Before the First World War, research results on turbulence were mainly published in the conference reports of academies which were then open to scientific and technological matters of all sorts:

Comptes Rendues de l'Académie des Sciences
Göttinger Nachrichten der Kgl. Gesellschaft der Wissenschaften
Journal and Proceedings of the Royal Society of New South Wales
Mathematische Abhandlungen der Königlichen Akademie der Wissenschaften
Nova Acta Regiae Societatis Scientiarum Upsaliensis
Philosophical Transactions of the Royal Society
Proceedings of the Royal Irish Academy
Proceedings of the Royal Society
Sitzungsberichte der Bayerischen Akademie der Wissenschaften
Sitzungsberichte der kaiserlichen Akademie der Wissenschaften zu Wien.

Otherwise investigations on turbulence from the nineteenth and early twentieth century are found in a host of journals. None of them, however, was dedicated to fluid mechanics only, not to speak of turbulence:

Annalen der Hydrographie und maritimen Meteorologie
Annales de Chimie et de Physique
Annales des mines
Annalen der Physik
Denkschrift des Verbandes deutscher Architekten- und Ingenieur-Vereine
Enzyklopädie der Mathematischen Wissenschaften
Forschungsarbeiten auf dem Gebiete des Ingenieurwesens
Handbuch der Mechanik fester Körper und der Hydraulik
Zeitschrift für Flugtechnik und Motorluftschiffahrt
Jahrbuch der Motorluftschiff-Studiengesellschaft
Jahresbericht der Deutschen Mathematiker-Vereinigung
Journal de Physique, de Chimie, d'Histoire Naturell et des Arts
Journal of Geography

Journal of the Association of Engineering Societies
Journal of the Scottish Meteorological Society
Meteorologische Zeitschrift
Philosophical Magazine
Physikalische Zeitschrift
Proceedings of the Royal Institution of Great Britain
Quarterly Journal of the Royal Meteorological Society
Report of the British Association for the Advancement of Science
Reports and Memoranda of the Committee for Aeronautics
Transactions of the American Society of Civil Engineering
Verhandlungen der Gesellschaft Deutscher Naturforscher und Ärzte
Verhandlungen des Internationalen Mathematiker-Kongresses
Zeitschrift für Mathematik und Physik.

In view of such heterogeneity it is not astonishing that there was no universal parlance about turbulence that could have discerned specific problems for closer study. The lack of a common stage is also reflected by mutual ignorance of prior achievements which anticipated one or another discovery. With regard to the riddle of the two manners of flow, laminar and turbulent, only Reynolds's paper from 1883 is celebrated as a landmark although it was anticipated by Hagen and others (see Sect. 1.6). In retrospect it is also striking that Reynolds introduced in his landmark paper of 1895 the averaged Navier–Stokes equations in order to cope with fully turbulent flows without any reference to Boussinesq's eddy viscosity concept. From a modern textbook perspective "Boussinesq's hypothesis" is introduced as if it constitutes a method of solving "Reynolds averages", although it preceded Reynolds' paper by more than two decades. "How is it possible that Boussinesq proposed in 1872 a closure for an equation that would be written more than 20 years later?", a study on these landmarks in the history of turbulence addressed this paradox. "The historical facts are somewhat more complex than what is usually mentioned in today's turbulence text books" (Schmitt, 2007, p. 619).

9.2 Scientific Engineering

Yet research on turbulence was not entirely arbitrary and scattered between diverse environments. Hydraulics, naval architecture and aeronautics came of age under an umbrella that may be labelled as scientific engineering. It was taught and cultivated in institutions like the École Polytechnique and the École des Ponts et Chaussées in France which served as role models for Polytechnic Schools and Technical Universities elsewhere. The institutionalization of scientific engineering in different countries followed different national traditions, but by the early twentieth century most industrialized countries disposed over institutions and facilities that could be subsumed under this label. Scientific engineering provided an early environment for research on turbulence before "turbulence" became its common designation. Reynolds, for example, was professor for Civil and Mechanical Engineering at the Owens College in Manchester (the future University of Manchester). In an obituary Reynolds was called Great Britain's "most distinguished scientific engineer" (Launder and Jackson, 2011, here p. 35).

In Germany, the tradition of scientific engineering was peculiar. During the first half of the nineteenth century it was growing at newly established Polytechnic Schools following the French role model of the École Polytechnique; however, by the end of the nineteenth century, when the Polytechnic Schools were transformed into Technical Universities (Technische Hochschulen), they had to struggle for equal rights with the Universities. Prandtl, for example, began his career at the Technical Universities in Munich and Hanover, before he rose to prominence at the University of Göttingen. At this university there was a strong tradition of mathematics and physics (initiated by Carl Friedrich Gauss (1777–1855) and Wilhelm Weber (1804–1891)) to which the entrepreneurial mathematician Felix Klein added applied sciences. When Prandtl was called to Göttingen he left no doubt about his "sense of belonging" to Technical Universities where he had wished to contribute to the "elevation of scientificity". Nevertheless he accepted the professorship for technical physics at Göttingen University because he hoped that "through the exchange of ideas with the theorists, I could help to solve some of the questions that are close to practice".[1] Thus, by the beginning of the twentieth century, in Germany too the tradition of scientific engineering had taken hold in an academic environment.

Scientific engineering, however, was not the only institutional environment for turbulence research. Problems like flow instabilities or vortex formation also attracted physicists and mathematicians with little interest in engineering. Prandtl's boundary layer concept was first presented at the Third International Congress of Mathematicians in August 1904 in Heidelberg (Prandtl, 1905). Sommerfeld presented his contribution to hydrodynamic stability theory four years later at the Fourth International Congress of Mathematicians in 1908 in Rome (Sommerfeld, 1909). Orr was professor of mathematics; he had presented his theory one year before to the Royal Irish Academy (Orr, 1907). Before it was staged under the label of "the turbulence problem" in *ZAMM*, the Orr–Sommerfeld approach belonged to the domain of theoretical physicists and mathematicians.

After World War I specialities such as aeronautics received an increased attention. Scholars like Lord Rayleigh or Prandtl assumed new roles as advisers for a new technology. In Germany, Prandtl suggested to create a "federation of all like-minded," as he wrote in a letter to von Mises in 1921. He had already discussed his plans with his former disciple Theodore von Kármán, who was now director of the Institute for

[1] "Das schwerste Bedenken entsprang meinem Zugehörigkeitsgefühl zur Technik. Es war seit langem ein Lieblingsgedanke von mir, nach Kräften an der Hebung der Wissenschaftlichkeit im Unterricht an den technischen Hochschulen mitzuarbeiten. Unter diesem Gesichtspunkt scheint mir der Übertritt an die Universität nur dadurch zu rechtfertigen zu sein, dass ich in dieser Stelle, die wie ich denke, nicht meine letzte sein wird, außerordentliche Gelegenheit haben würde, meinen eigenen wissenschaftlichen Wert zu heben und mich für künftige Aufgaben vorzubereiten, andererseits auch durch den Gedankenaustausch mit den Theoretikern noch manche der Praxis naheliegende Frage lösen helfen könnte." Quoted in Eckert (2017, p. 45).

Aerodynamics at the Technische Hochschule Aachen. They proposed the foundation of an "Association for Technical Mechanics" with the exclusive purpose to prepare and convene meetings for that speciality. The audience would be mainly "scientific engineers".[2] Mises addressed the same group as editor of the new *Zeitschrift für Angewandte Mathematik und Mechanik (ZAMM)*. In his editorial he discerned turbulence among other problems as a particular challenge. "According to the present state of theory, it must be considered undecided whether the approach of viscous fluids, with sufficient mathematical penetration, is able to provide an explanation of turbulence."[3] Besides applied mathematics, technical or applied mechanics was perceived as a discipline of its own right. Prandtl and Mises did not agree entirely about the label of the new organization. It was founded in 1922 as Gesellschaft für Angewandte Mathematik und Mechanik (GAMM), against Prandtl's and Kármán's preference for "technical" mechanics. But they shared the concern about turbulence as an outstanding problem that deserved particular attention.

The first occasion to present their goals to broader circles of like-minded engineers, physicists and mathematicians came in September 1921, when the Deutsche Physikalische Gesellschaft, the Deutsche Gesellschaft für Technische Physik and the Deutsche Mathematiker-Vereinigung combined their annual meetings in a common conference in order to bring to bear the areas of applied mathematics and mechanics to a higher degree than heretofore, as Mises reported in *ZAMM* (von Mises, 1921b). Prandtl used the occasion for "Remarks about the Onset of Turbulence" (Prandtl, 1921); Schiller surveyed the experimental investigations about the onset of turbulence (Schiller, 1921). Noether reviewed "the turbulence problem" from a theoretical perspective (Noether, 1921). This paper served as a starting point for a host of further investigations of the Orr–Sommerfeld approach (see Chap. 8).

9.3 The International Congresses for Applied Mechanics

The initiative to form a community of applied scientists soon became international. In 1922, Kármán (by birth Hungarian) and the Italian mathematician Tullio Levi-Civita (1873–1941) organized a gathering of like-minded mathematicians, physicists and engineers with an interest in fluid mechanics at Innsbruck in Austria (Battimelli,

---

[2] "… haben in mir den Gedanken enstehen lassen, dass es sehr wünschenswert wäre, für das Gebiet der 'technischen Mechanik' (im weitesten Sinn genommen) auf den Mathematiker- und Physikertagen durch einen Zusammenschluss aller Gleichstrebenden einen festen Sammelpunkt zu schaffen. Ich habe mit Kármán darüber gesprochen und seinen vollen Beifall bekommen. Wir denken an die Gründung einer 'Vereinigung für technische Mechanik', deren einziger Zweck die Vorbereitung und Abhaltung von fachwissenschaftlichen Versammlungen sein soll. […] weil mir vor allem daran liegt, die wissenschaftlichen Ingenieure zusammen zu bekommen …" Prandtl to Mises, 2 August 1921. MPGA, Abt. III, Rep. 61, Nr. 1078.

[3] "Nach dem gegenwärtigen Stand der Theorie muß man es als noch unentschieden ansehen, ob der Ansatz der zähen Flüssigkeiten bei genügender mathematischer Durchdringung eine Erklärung der Turbulenz zu geben vermag" (von Mises, 1921a, p. 12). On Mises' editorial goals with *ZAMM* see Siegmund-Schultze (2020).

1996; von Kármán and Levi-Civita, 1924). In the following year the momentum of these events was carried on to the Netherlands, where Burgers and Cornelis Benjamin Biezeno (1888–1975) organized the First International Congress for Applied Mechanics, held in April 1924 at Delft (Alkemade, 1995; Battimelli, 1988). Two years later, the Second International Congresses for Applied Mechanics was held in Zurich, Switzerland. Thereafter, they were held every four years in another country (1930 Stockholm; 1934 Cambridge, UK; 1938 Cambridge, Massachusetts, USA). The Second World War interrupted this sequence until it was resumed under the new umbrella of the International Union of Theoretical and Applied Mechanics (IUTAM) (Doel et al., 2005; Eberhard and Juhasz, 2016) (see Sect. 11.2).

From the very beginning, these congresses were regarded as the preferred stage for presenting novel research on turbulence. Their importance can hardly be exaggerated. At each one of these meetings crucial achievements were introduced and discussed. A few examples may suffice. At Innsbruck, Kármán presented his most recent theory on turbulent skin friction (von Kármán, 1921, 1922); Heisenberg, Sommerfeld's prodigy student, reported about some results of his early struggle with the turbulence problem that subsequently became the subject of his doctoral work (Heisenberg, 1922, 1924); at the Zurich Congress in 1926, Prandtl presented the mixing length concept (Prandtl, 1927); four years later Kármán chose the Stockholm Congress to introduce the similarity concept that gave rise to "Kármán's constant" and the logarithmic "wall law" for turbulent skin friction (von Kármán, 1930). As Kármán recalled in his autobiography, these Congresses served as the "playing field" for his competition with Prandtl. "The 'ball' was the search for a universal law of turbulence" (von Kármán, 1967, p. 134).

At the Fourth International Congress for Applied Mechanics in Cambridge, UK, the interest in turbulence was so great that the organizers of the subsequent mechanics congress decided to convene a special symposium on turbulence within that event. "Professor Prandtl kindly consented to organize this Symposium," the organizers reported in the proceedings of the Fifth International Congress for Applied Mechanics, held in Cambridge, Massachusetts, USA. They regarded this Symposium "not only the principal feature of this Congress, but perhaps the Congress activity that will materially affect the orientation of future research" (Hartog and Peters, 1939). Taylor's statistical theory of turbulence had been published just three years ago; his survey lecture (Taylor, 1939), Dryden's report on experimental measurements of the spectrum of turbulence at the National Bureau of Standards in Washington D.C. (Dryden, 1939) and other contributions at this symposium illustrated that the focus had shifted from the onset of turbulence to the statistical theory of fully developed turbulence.

One reason for this shift was the improvement of hot-wire anemometry and a growing concern about wind tunnel turbulence. "The desire to make some direct measurement of the turbulence in wind tunnels was the incentive for the work here described," Dryden and a collaborator had argued already ten years before in a NACA Report. They were optimistic that hot-wire anemometry also advanced the statistical theory of fully developed turbulence, because the "great need at the present time in the further development of the theory of turbulence is more experiments on the actual fluctuations to supplement the data already available on the distribution of mean velocity" (Dryden and Kuethe, 1929b, p. 361). In a subsequent report they concluded "that turbulence is a variable of some importance at all times and that the careful experimenter will desire to measure and state its value in order that his experiments may be capable of interpretation" (Dryden and Kuethe, 1929a, p. 166). The symposium on turbulence at the Fifth International Congress on Applied Mechanics served as an ideal stage for presenting the progress made in this area during the preceding decade—achieved mainly by Dryden and his group at the National Bureau of Standards in cooperation with NACA.

While the statistical theory of turbulence rose to prominence as a tool for the analysis of wind tunnel turbulence, the onset of turbulence was not completely set apart. However, it was not perceived as the result of an instability. Taylor exposed at the turbulence symposium in great detail how the onset of turbulence in the boundary layer of an object in a wind tunnel depends on the degree of turbulence in the air-stream. Given the apparent role of external disturbances from the air-stream as a cause for the transition to turbulence, an intrinsic boundary layer instability as analysed by Tollmien and Schlichting seemed an unlikely mechanism for the onset of turbulence. "It seems to me not only that it is not proved that the boundary layer is unstable," Taylor concluded his survey, but that also "the experimental evidence is against the instability theory" (Taylor, 1939, p. 308). Yet stability theory was not entirely absent at the turbulence symposium. Synge lectured on "The Stability of Plane Poiseuille Motion" (Synge, 1939)—after he had surveyed hydrodynamic stability more broadly already at the occasion of the fiftieth birthday of the American Mathematical Society (Synge, 1938) (see Sect. 8.3).

Thus the stage was set for turbulence as an international research field under the label of applied mathematics and mechanics—but the outbreak of World War II prevented further exchanges of research results.

Chapter 10
Turbulence in WW II

Abstract In World War II research on turbulence became part of secret war projects. The development of wings with minimal drag ("laminar wings") involved investigations about the transition to turbulence in the boundary layer. Fully developed turbulence was investigated because of its impact on skin friction, heat transfer and other practical applications. But progress was also achieved in more fundamental areas with little or no relation to wartime projects. Research on turbulence in WW II, therefore, manifested itself on a variety of stages. The communication of research results was mostly—but not always—confined to secret or confidential wartime reports. More fundamental contributions (such as K41 and similar work of Prandtl) were published without restrictions despite the war.

10.1 Wartime Reports

The symposium on turbulence during the Fifth International Congress for Applied Mechanics in September 1938 represented the last meeting before World War II at which research results on turbulence were freely communicated on an international stage. The spectre of war was already present at this occasion with the imminent German invasion of Czechoslovakia that led to controversial political debates among the participants of the congress. "I realized that you know nothing of what the criminal lunatic who rules your country has been doing," Taylor wrote to Prandtl after the congress.[1] The correspondence between these two outstanding representatives of international research on turbulence came to an end when Prandtl defended Nazi politics (Eckert, 2017, Chap. 7). It caused the break of their personal relation as well as the end of the unrestricted exchange of research results even before the outbreak of war.

On a national scale, of course, turbulence research did not come to a halt. The stages for the communication of results, however, changed with the outbreak of war. Applied mechanics, and fluid mechanics in particular, was pertinent to a host of war technologies, from submarines to airplanes, bombs and rockets. In many cases the development of advanced weapons involved turbulence. A great deal of research on

[1] Taylor to Prandtl, 27 September 1938. AMPG, Abt. III, Rep. 61, No. 1654.

M. Eckert, *Turbulence—an Odyssey*, History of Physics,
https://doi.org/10.1007/978-3-030-91459-2_10

turbulence during World War II was performed in the context of one or another war project. A few examples may illustrate this kind of research projects.

One of these concerned the investigation of turbulent jets. It was performed under NACA contract at the GALCIT and carried out by Kármán's doctoral student Stanley Corrsin (1920–1986) (1943) (Fig. 10.1). In this investigation Corrsin employed a novel hot-wire measuring set. Thus began the career of an expert on turbulence who would rise to prominence after the war (Meneveau and Riley, 2011).

Another disciple of Kármán who became involved in turbulence investigations during the war was Liepmann (1943). His research was pertinent for a larger NACA program concerned with laminar wings and low-turbulence wind tunnels (Hansen, 1987, pp. 109–116; Roland, 1985, vol. 2, p. 549). The NACA issued similar research contracts also to Dryden's Aerodynamics Division of the National Bureau of Standards in Washington, D.C. In the course of these investigations, performed in special low-turbulence tunnels with sophisticated hot-wire equipment, both Dryden's and Kármán's groups discovered the boundary layer instability predicted by Tollmien and Schlichting at Göttingen (see Sect. 8.4). The way of presenting this discovery to those with a need to know was in the form of a war report issued in April 1943 (Schubauer and Skramstad, 1943); it was made public after the war as "Report No. 909" on the stage of NACA reports—by then a widely renowned outlet for communications on aeronautical research.

In Germany, the Tollmien–Schlichting theory had been further pursued during the war with the goal to control the onset of turbulence in the boundary layer. In 1941 Schlichting reviewed the theory from the perspective of laminar wings at a seminar in a secret aeronautical research facility near Braunschweig, the Luftfahrtforschungsanstalt Hermann Göring, LFA, to which his institute of aerodynamics at the Technical University Braunschweig was associated. His lecture was subsequently distributed in the form of a secret war report (Fig. 10.2).

Schlichting was also well informed about laminar wing research in the USA and in Japan (Schlichting, 1941, p. 30; Schlichting and Bußmann, 1942, p. 17). In 1943 two collaborators of Schlichting's institute reported on "Measurements of the original wing of the type P-51 Mustang" that was known "for some time" for its laminar profile (Breford and Möller, 1943; Kopfermann and Breford, 1943). Another wartime report evaluated the properties of "Russian laminar profiles" (Riegels, 1943). During these investigations it became clear how much the property of laminar flow in the boundary layer depends on the smoothness of the wing surface. Without low-turbulence tunnels, however, laminar wing investigations remained inconclusive. The Germans built such tunnels, but they did not become operative before the end of the war (Holstein, 1946).

Laminar wings were not the only wartime applications that involved turbulence. Prandtl's close collaborator in all matters of turbulence, Karl Wieghardt (1913–1996), conducted a series of experiments in a "roughness tunnel" ("Rauhigkeitskanal"), a special wind tunnel with a rectangular cross section where one tunnel wall was composed of movable parts so that a prescribed field of pressure could be generated over the surface of an inserted test model. This wind tunnel was part of Prandtl's research program on turbulence since the mid-1930s. In 1940, additional equipment

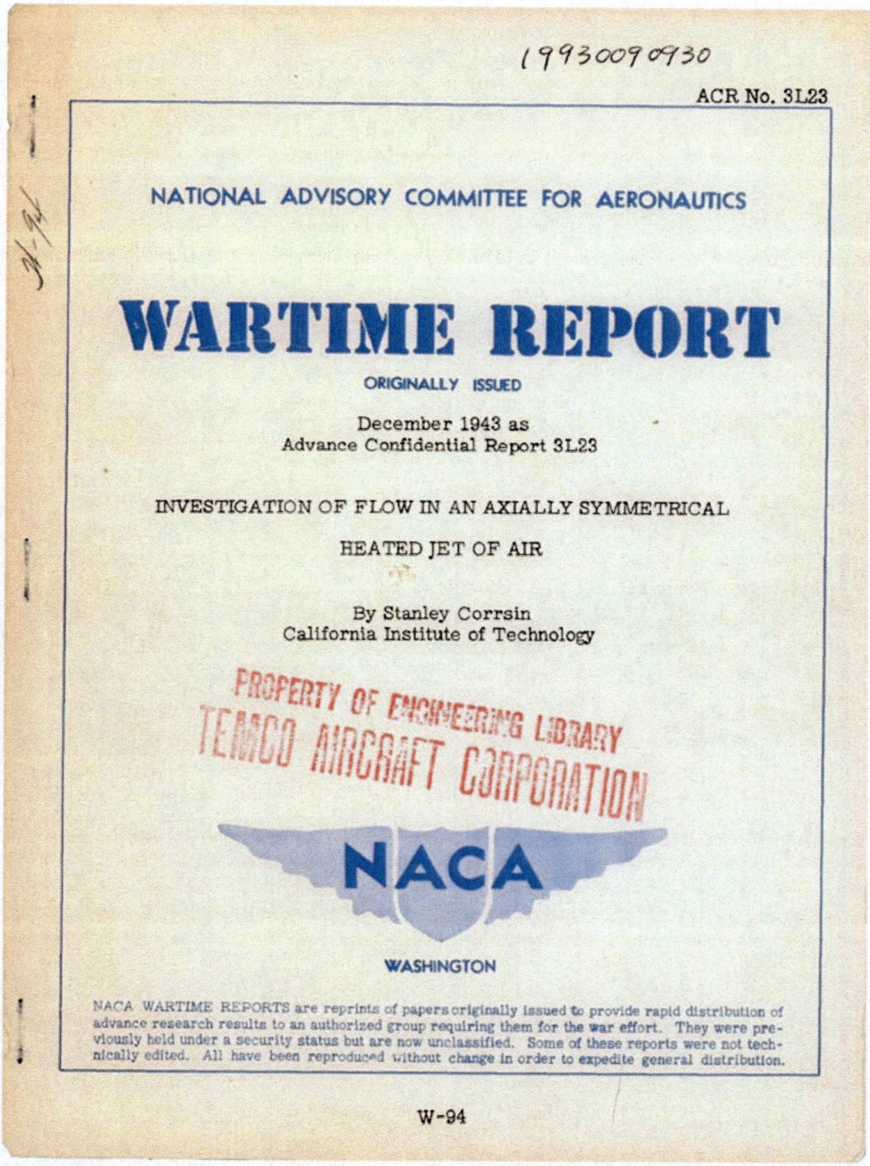

19930090930

ACR No. 3L23

NATIONAL ADVISORY COMMITTEE FOR AERONAUTICS

WARTIME REPORT

ORIGINALLY ISSUED

December 1943 as
Advance Confidential Report 3L23

INVESTIGATION OF FLOW IN AN AXIALLY SYMMETRICAL

HEATED JET OF AIR

By Stanley Corrsin
California Institute of Technology

PROPERTY OF ENGINEERING LIBRARY
TEMCO AIRCRAFT CORPORATION

NACA

WASHINGTON

NACA WARTIME REPORTS are reprints of papers originally issued to provide rapid distribution of advance research results to an authorized group requiring them for the war effort. They were previously held under a security status but are now unclassified. Some of these reports were not technically edited. All have been reproduced without change in order to expedite general distribution.

W-94

Fig. 10.1 Title page of Corrsin's report on turbulence in jets (Corrsin, 1943)

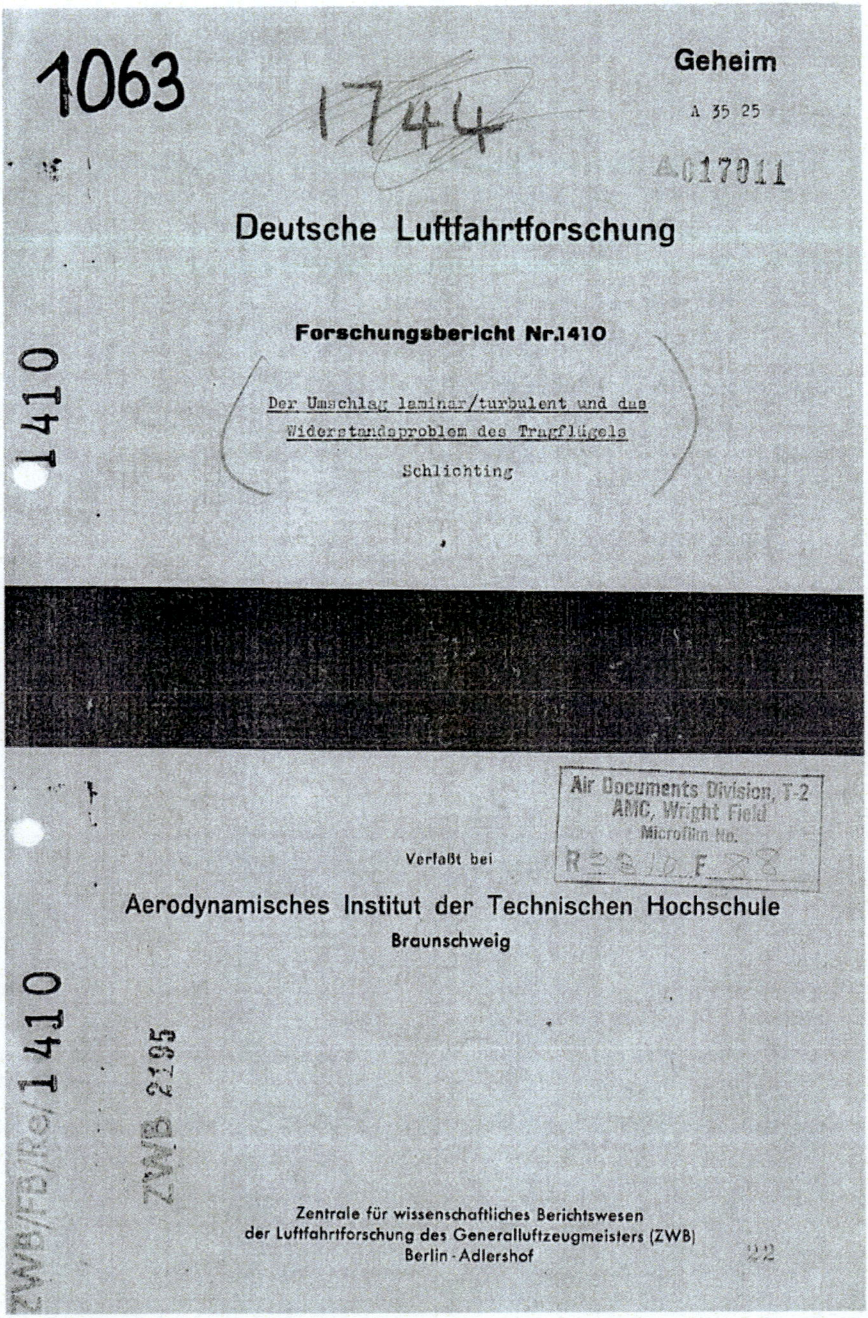
1063

1744

Geheim

A 35 25

A017811

Deutsche Luftfahrtforschung

Forschungsbericht Nr.1410

Der Umschlag laminar/turbulent und das Widerstandsproblem des Tragflügels

Schlichting

1410

[redacted]

Air Documents Division, T-2
AMC, Wright Field
Microfilm No.

Verfaßt bei

Aerodynamisches Institut der Technischen Hochschule
Braunschweig

ZWB/FB/Re/1410

ZWB 2195

Zentrale für wissenschaftliches Berichtswesen
der Luftfahrtforschung des Generalluftzeugmeisters (ZWB)
Berlin-Adlershof

Fig. 10.2 Title page of Schlichting's war report on "Transition from laminar to turbulent flow and the drag problem of air foils" (Schlichting, 1941) was installed in order to use it for war projects such as skin friction measurements of rubber linings to shield submarines from acoustic detection (Wieghardt, 1944). Another war research focussed on the "spreading effects in turbulent friction layers", as Prandtl and Wieghardt titled this report in September 1944. It was concerned with the spreading "of warfare agents or artificial fog". As it was not possible to experiment with poison gas in the laboratory, Wieghardt resorted to the analogy between the turbulent diffusion of gas and the spreading of heat in an air current. Thus he measured the distribution of temperature in the turbulent wake of a heat source placed inside the roughness tunnel. The results were displayed so that the spreading of gas from a source could be quickly read as a function of different wind speeds.[2]

These and dozens of other wartime reports indicate a broad range of applications that involved turbulence. The variety of applications makes it difficult to categorize this research. In a postwar review Prandtl distinguished between "Wall-bounded turbulence", "Free turbulence" and "Individual problems" (Prandtl, 1953, p. 55). Another report of German wartime research on turbulence lists these topics: "Stability theory of laminar flows", "Turbulent friction layers", "Free turbulence", "A series of fundamental investigations" and "Geophysical applications" (Görtler, 1948, p. 75). The *Monographien über Fortschritte der deutschen Luftfahrtforschung (seit 1939)*, solicited by the British military government and edited by the director of the Aerodynamische Versuchsanstalt (AVA) in Göttingen, Albert Betz (1885–1968), reviewed research on turbulence in a volume on "Boundary layers" under the headline "Turbulent friction layers". The subjects addressed in this manner dealt not only with skin friction but also with other boundary layer phenomena like heat transfer between a hot solid body and the surrounding flow.

War research on turbulence, therefore, could appear under different umbrellas and on various stages. As Kármán had observed few years before the war, "if we meet a practical question in aerodynamic design which we are unable to answer, the reason that we are unable to give a definitive answer is almost certainly that it involves turbulence" (von Kármán, 1937b, p. 1141).

10.2 Fundamental Research on Turbulence During the War

But turbulence research in World War II was not entirely confined to one or another war project. Some results, such as the instability along concave walls ("Görtler instability") were even published in the open literature (Görtler, 1940a, b, 1941). The more fundamental a research appeared the less it was restricted to the stage of war reports. But there is no sharp boundary between both categories. What appears fundamental in one context may be considered as applied in a different context. In the same

[2] K. Wieghardt: Über Ausbreitungsvorgänge in turbulenten Reibungsschichten. Geheimbericht für das Marineobservatorium Greifswald, 1. September 1944. APMG, Abt. III, Rep. 76B, Kasten 2. For more detail see Schmaltz (2005, pp. 340–352).

vein applications may encourage more fundamental considerations, sometimes even on the same stage. Without closer inspection of the context, the attributes "basic" versus "applied" hardly provide a useful categorisation of research. Both terms are associated with political and ideological meanings that make an unbiased assessment difficult (Kaldewey and Schauz, 2018).

A few examples may suffice to illustrate the blurring of boundaries between basic and applied research. In the context of laminar airfoils the Tollmien–Schlichting theory appears as a prime case of applied science. But its pertinence for aeronautical applications also evoked broader interest in the underlying principles. The conclusion of Schlichting's seminar on the laminar-turbulent boundary layer transition is telling:

> Finally I would like to conclude that the problem of the onset of turbulence can in principle be considered solved from the theoretical side. The method of small oscillations provides satisfactory results if the physically important quantities of profile curvature and viscosity are taken into account in the right way. The first solution according to this method dealt with the problem of plate flow (plane wall, no pressure gradient, Blasius profile). But also the problem of calculating the transition point for an airfoil, which is important for aeronautical engineering, has been solved in principle, as I have shown with the examples given.[3]

Therefore Schlichting made boundary layer theory in winter 1941/42 the subject of a lecture series. The site was once more the secret Luftfahrtforschungsanstalt Hermann Göring (LFA). Yet these lectures were perceived as fundamental textbook knowledge for engineers. Schlichting's lecture notes (Schlichting, 1942) were translated and published after the war as NACA Technical Memoranda on "Laminar Flows" and "Turbulent Flows" (Schlichting, 1949a, b) (Fig. 10.3). Thus the stage was set for the first textbook on *Boundary-Layer Theory* (Schlichting, 1951). It went through many editions and became an engineering textbook classic that is still used today (Schlichting and Gersten, 2017).

Fully developed turbulence was also subject of fundamental research during the war, such as Kolmogorov's theory. At first sight, K41 may seem a by-product of Kolmogorov's applied war-work as a statistician: when the German troops invaded Russia in 1941, Kolmogorov suggested a stochastic distribution of barrage balloons over Moscow as a protection against bombing (Chentsov, 1990, p. 987). But a closer look at the prehistory of K41 shows that this theory was unrelated to Kolmogorov's war work. It was rather a response to the investigations of wind tunnel turbulence and the ensuing statistical theories of Taylor and von Kármán from the late 1930s (see Sect. 7.5).

The same holds for Prandtl. Despite his involvement in wartime projects he did not completely abstain from pursuing his pre-war agenda. In autumn 1944 he elaborated some of the ideas which he had introduced in 1938 on the stage of the turbulence

[3] "Abschließend möchte ich feststellen, dass das Problem der Turbulenzentstehung von der theoretischen Seite prinzipiell als gelöst gelten kann. Die Methode der kleinen Schwingungen liefert befriedigende Ergebnisse, wenn man die physikalisch wichtigen Größen Profilkrümmung und Zähigkeit in richtiger Weise berücksichtigt. Zuerst gelöst wurde nach dieser Methode das Problem der Plattenströmung (ebene Wand, kein Druckgradient, Blasius-Profil). Aber auch das für die Flugtechnik wichtige Problem der Berechnung des Umschlagpunktes für ein Tragflächenprofil ist prinzipiell gelöst, wie ich mit den angegebenen Beispielen gezeigt habe." (Schlichting, 1941, p. 50).

NACA TM No. 1218

7879

NATIONAL ADVISORY COMMITTEE
FOR AERONAUTICS

TECHNICAL MEMORANDUM

No. 1218

LECTURE SERIES "BOUNDARY LAYER THEORY"

PART II - TURBULENT FLOWS

By H. Schlichting

Translation of "Vortragsreihe" W.S. 1941/42, Luftfahrtforschungsanstalt Hermann Göring, Braunschweig

NACA

Washington
April 1949

Fig. 10.3 Title page of Schlichting's lecture series that set the stage for his textbook on boundary-layer theory (Schlichting, 1949b)

symposium of the mechanics congress in Cambridge, Massachusetts. His focus was similar to that of Kolmogorov: the energy per unit volume contained in the turbulent fluctuations. His unpublished notes show that he aimed in 1944/1945 at a general theory of developed turbulence. In this venture he also arrived at the result for the downgrading of turbulent motions to what later was labelled "Kolmogorov's micro-scale" (Bodenschatz and Eckert, 2011, Sect. 2.10). On 4 January 1945 Prandtl presented preliminary results at an internal colloquium to the collaborators of his institute. Three weeks later Prandtl and Wieghardt submitted a paper “On a new system of equations for fully developed turbulence” (“Über ein neues Formelsystem für die ausgebildete Turbulenz”) to the Göttingen Academy of Science (Prandtl and Wieghardt, 1945). Despite the war, this contribution to the theory of turbulence appeared without restrictions in the proceedings of this academy.

Chapter 11
Postwar Revelation and Reorganization

Abstract The revelation of wartime achievements fuelled turbulence research in many countries. The rehabilitation of the Orr–Sommerfeld approach encouraged new investigations about the onset of turbulence. The theory of fully developed turbulence as elaborated by Kolmogorov and others became a fashionable research theme. The International Union of Theoretical and Applied Mechanics (IUTAM) provided a new international stage for presenting achievements of turbulence research. In the USA turbulence became an integral part of a new Division of Fluid Dynamics (DFD) under the umbrella of the American Physical Society (APS).

11.1 Disclosing Wartime Achievements

Like the preceding 1938 mechanics conference, the Sixth International Congress for Applied Mechanics, held from 22 to 29 September 1946 in Paris at the Sorbonne, became an important stage for presenting turbulence research. So shortly after the war Germany and Austria were absent, but otherwise it was a truly international event, attended by scientists from France, the USA, England, Belgium, Holland, Italy, Switzerland, Romania, Czechoslovakia, Sweden, Turkey, Poland, Russia and China. Once more a special symposium on turbulence was organized. Dryden introduced this symposium with a report on "Some Recent Contributions to the Study of Transition and Turbulent Boundary Layers"—disclosing the wartime investigations of his group at the National Bureau of Standards that resulted in the discovery of Tollmien–Schlichting waves. As Dryden concluded in *Science*:

> The paper aroused much interest, since it contained the first public announcement of the work of Schubauer and Skramstad, both of the National Bureau of Standards, in verifying experimentally the Tollmien-Schlichting theory of boundary layer instability and the first complete measurements of the turbulent stress tensor in a separating turbulent boundary layer. (Dryden, 1947, p. 167)

The experimental discovery of Tollmien–Schlichting waves entailed debates about stability theory. When Kármán presented Lin's review of the Orr–Sommerfeld approach he encountered the objection of a mathematician from King's College in London:

M. Eckert, *Turbulence—an Odyssey*, History of Physics,
https://doi.org/10.1007/978-3-030-91459-2_11

> D. Meksyn, who presented a paper on the stability of viscous motion between parallel planes, was sharply critical of the work of Tollmien and others. A full report of his work is to be published in the Proceedings of the London Mathematical Society. He seemed to have made some improvement in the accuracy of the mathematical computation of the stability limit but insisted that the other workers, including Lin, were wrong in spite of the good experimental confirmation of their work by the experiments of Schubauer and Skramstad at the National Bureau of Standards. (Dryden, 1947, p. 168)

Thus Dryden's revelation at Paris did not settle the dispute about the onset of turbulence but signalled a new round of controversial debates. Meksyn agreed that the stability equation of the Orr–Sommerfeld approach "is of very great importance in the problem of turbulent motion" but he regarded Heisenberg's and Tollmien's theories as "incomplete" because they failed to develop approximations for "rapidly varying integrals" (Mekysn, 1946, pp. 436–437). Hence transition to turbulence remained a challenge for mathematicians even in the simplest cases of two-dimensional plane flows.

Another revelation at the Paris congress concerned Kolmogorov's statistical theory of turbulence (see Sect. 7.5). It had been unknown outside the USSR although Kolmogorov's papers were published in 1941 in the English-language edition of the Proceedings of the Sovjet Academy of Science. Taylor's doctoral student Batchelor had come across these papers in the Cambridge library and made them part of a brief note. He also revealed that Kolmogorov's theory was not the only achievement in this regard. Some of Kolmogorov's results were obtained in a "remarkable series of coincidences" also by Lars Onsager (1903–1976), Carl Friedrich von Weizsäcker (1912–2007) and Werner Heisenberg. Onsager had shown Batchelor an account of his work prior to publication. Weizsäcker and Heisenberg had shown their unpublished manuscripts during their detention to Taylor. Their results "will probably be described in a German publication" (Batchelor, 1946, p. 884).

Many years later Batchelor recalled how he first encountered this theory which became the main subject of his further career. He had arrived in Cambridge in April 1945 in order to work for a Ph.D. on turbulence under Taylor's supervision. Kolmogorov's papers that he found in the Cambridge library marked the beginning of his studies. "I was excited by them, and told G.I. what I had found." A few months later, in summer 1945, Heisenberg and Weizsäcker visited Taylor in order to discuss their ideas on turbulence that they were developing during their idle time in England as detainees in Farm Hall near Cambridge. They described "some new ideas on the spectral representation of isotropic turbulence and the transfer of energy from large to small-scale components, and although these ideas were embedded in unfamiliar analysis it was evident to G.I. and me that they had points in common with Kolmogorov's theory" (Batchelor, 1996, pp. 170–171). Onsager's approach, by that time, could only be guessed from the abstract of a talk on "The Distribution of Energy in Turbulence" that he had presented in November 1945 at a meeting of the Metropolitan Section of the American Physical Society at Columbia University in New York (Onsager, 1945).

Batchelor would have been surprised to learn that this "remarkable series of coincidences" was not exhausted by the contributions of Kolmogorov, Onsager, Weizsäcker and Heisenberg. Prandtl, too, had arrived independently at the same results. His effort began in autumn 1944 and was pursued until summer 1945. It was his favourite subject of research during the last months of the war (Bodenschatz and Eckert, 2011).

The stages on which these contributions were developed and communicated were set by the wartime and early postwar circumstances. Kolmogorov and his collaborators published their papers in academy proceedings; for the "Russian school" turbulence became a long-time concern (Falkovich, 2011). The same holds for Prandtl. He perceived turbulence as a major challenge—both for his own research and for his collaborators at the Kaiser-Wilhelm-Institute. His ideas on fully developed turbulence in 1945 were published in the proceedings of the Göttingen academy (Prandtl and Wieghardt, 1945). In an unpublished paper "on the role of viscosity in the mechanism of developed turbulence" he developed the theory further, but refrained from publication when he learned that Weizsäcker and Heisenberg had arrived at a more advanced theory. He contented himself with mentioning it in a report to the Field Intelligence Agency, Technical, (FIAT), which reviewed German achievements in science during the war period (Prandtl, 1953, p. 77). Unlike Prandtl and Kolmogorov, Onsager's contribution to the theory of turbulence was a singular effort. He exchanged his ideas only by correspondence with some colleagues but never returned to the theory of turbulence after the publication of an elaborated version of his presentation at Columbia University (Eyink and Sreenivasan, 2006).

The circumstances under which Weizsäcker and Heisenberg approached the theory of turbulence in summer 1945 were entirely different. They belonged to the group of ten detainees from the "Uranium Club", the German nuclear wartime project. In order to make good use of the time of their arrest, they started a series of colloquia in which they reviewed their mutual fields of interest. On Friday, 25 May 1945, one of them (Erich Bagge, 1912–1996) entered in his diary: "This day was dedicated to a colloquium on turbulence."[1] The speaker was Weizsäcker. He was interested in the role of turbulence in the formation of the solar system (von Weizsäcker, 1943). At Farm Hall, he derived from dimensional considerations the scaling laws of the K41 theory. After their release Weizsäcker and Heisenberg further elaborated the theory. Among themselves, Heisenberg attributed the priority to Weizsäcker for deriving the scaling laws from similarity considerations, in particular the energy spectrum for high Reynolds numbers in the inertial range $E(k) \sim k^{-5/3}$, where k is the wave number. Heisenberg's part was to provide a consistent mathematical analysis in Fourier space which yielded also the spectrum for higher wave numbers beyond the inertial range $E(k) \sim k^{-7}$ (see Fig. 11.1). They submitted their papers by December 1946 to the *Zeitschrift für Physik*, but postwar circumstances delayed publication for almost two years (von Weizsäcker, 1948; Heisenberg, 1948).

Batchelor's report at the Paris mechanics congress in 1946 about this "remarkable series of coincidences" in the theory of isotropic turbulence and Dryden's disclosure of the experimental verification of the Tollmien–Schlichting instability were not the only revelations which made this meeting so remarkable. Dryden singled out

[1] "Der heutige Tag stand im Zeichen eines Kolloquiums über Turbulenztheorie" (Bagge et al., 1957, pp. 48–53).

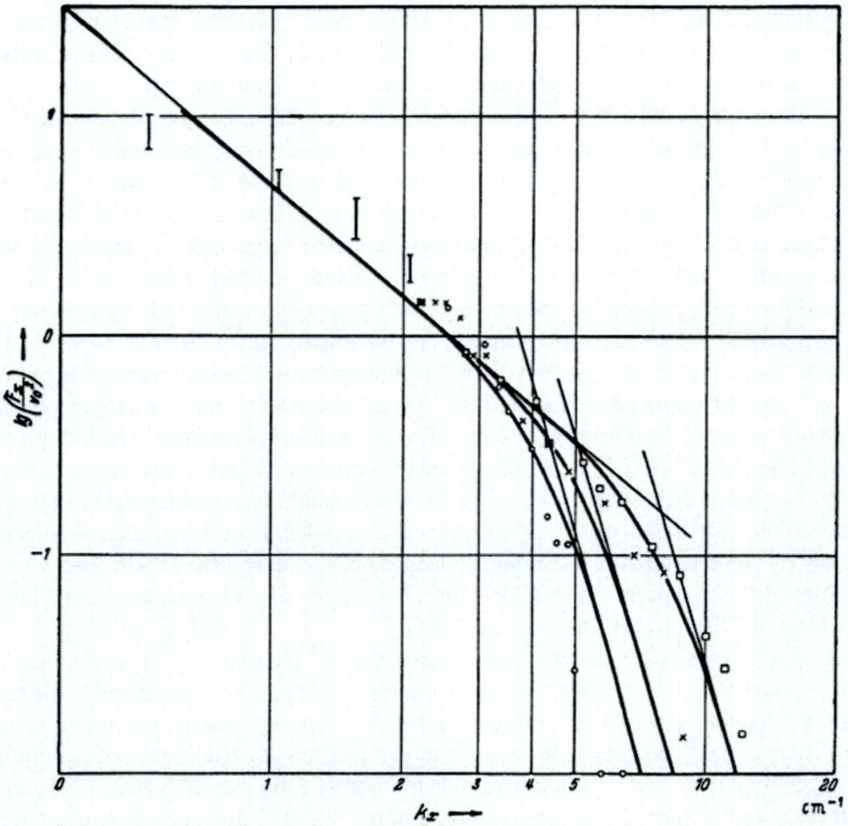

Fig. 11.1 The turbulent energy distribution as a function of the wave-number k, as derived by Heisenberg in 1946 and compared with then available data from grid-generated wind tunnel turbulence. The spectrum falls off proportional to $k^{-5/3}$ in the inertial range and to k^{-7} at smaller scales where the viscosity plays the predominant role. However, the Reynolds number in the experiments was not high enough so that "the range of validity of the $k^{-5/3}$-law is here so small that a reliable check is not possible" (Heisenberg, 1948, pp. 638–639). The $k^{-5/3}$-law was confirmed only much later (see Sect. 13.3)

several other contributions to the Paris turbulence symposium: Oscar Hinze's (1907–1993) investigation of high-speed liquid jets; Joseph Kampé de Fériet's (1893–1982) critique of Reynolds averaging; Alexandre Favre's (1911–2005) description of an apparatus for the measurement of time correlations in turbulent flow; Francois Naftali Frenkiel's (1910–1986) turbulence measurements conducted in his doctoral work under the supervision of Kampé de Fériet (Dryden, 1947). The proceedings have never been published, but there can be no doubt that this congress put turbulence on the stage for future discussions.

11.2 IUTAM

The Paris congress was part of the international postwar reconstructions on a larger scale. Although the International Congresses for Applied Mechanics were organized by international congress committees, prior to the Second World War there was no institutional umbrella in the form of an official international organization. The initiative to lay down a more permanent international basis for mechanics came from Burgers. Since the 1920s he and von Kármán performed the role as missionaries for international science. Now they envisioned a Union of Applied Mechanics under the umbrella of the International Council of Scientific Unions (ICSU), which had served already before the war as an organizational framework for other disciplines. The Paris conference provided the stage for further discussions. On 26 December 1946, the International Union of Theoretical and Applied Mechanics (IUTAM) was officially constituted; in April 1947, its statutes were approved, and in September 1947, a provisional bureau was set up, with Dryden as acting treasurer, and Burgers as acting secretary (Schiehlen and Wijngaarden, 2000, p. 47). With regard to politics and funding, IUTAM, as part of ICSU, was closely related to the United Nations Educational, Scientific and Cultural Organization (UNESCO), another newly established framework for international collaboration after the Second World War (Greenaway, 1996; Hult and Hoff, 2016).

Among IUTAM's first activities was the organization of the Seventh International Congress for Applied Mechanics. It was scheduled to take place in September 1948 in London—two years earlier than according to the accustomed four year interval because otherwise it would have interfered with the International Congress of Mathematicians scheduled for 1950, another stage for presenting turbulence research. The London Congress confirmed the great interest in renewing the prewar tradition. Among the 882 participants the only Germans were those who had emigrated to other countries before the war. However, the absence of Prandtl and other German scientists hardly skewed the representation of mechanics as an international discipline. The London congress made visible to what extent the field had grown since the 1920s, in particular turbulence. The proceedings comprised four volumes, and turbulence was again an outstanding theme. Batchelor presented the General Lecture on "Recent Developments in Turbulence Research". Other remarkable presentations on turbulence were those on the decay of turbulence by Frenkiel ("Comparison between theoretical and experimental results of the decay of turbulence"), Lin ("On the law of decay and the spectrum of isotropic turbulence") and Townsend ("Diffusion in the turbulent wake of a cylinder"). A number of those who had presented papers on turbulence at Paris were again performing at London, like Batchelor, Dryden, Favre and Hinze. The London Congress in 1948 was another stage at which turbulence became visible as a research field in its own right (IUTAM, 1948). The subsequent quadrennial Congresses at Istanbul (1952), Brussels (1956) and Stresa (1960) confirmed IUTAM's role as an organizatorial framework that encouraged this development.

But Burgers had broader ambitions than offering a new umbrella for the mechanics congresses. He regarded it unsatisfactory that IUTAM would merely organize these four-year events and remain inactive in the interim period. His primary motive for the foundation of IUTAM was to enable cross-fertilization by cooperation with other Unions under the umbrella of ICSU. In order to encourage such cooperation, Burgers proposed to organize international symposia between the congresses and to establish committees whose task was to plan these events (Schiehlen and Wijngaarden, 2000, p. 49). While these committees consisted of individual members, IUTAM was formed by national scientific organizations as members. In 1948, these were only the Royal Society of London and the Hungarian Academy of Sciences. Germany was officially invited to join the ICSU in 1950. In the same year, GAMM became a member of IUTAM. Over the years, IUTAM grew into a truly international body of broad scope that represented a heterogeneous mix of different national research cultures (Juhasz, 1988, p. 16).

The year 1950 may be regarded as a watershed for the reception of German science in the international arena after World War II. In the same year German mathematicians were invited to participate in the International Congress of Mathematicians held in September 1950 in Cambridge, Massachusetts. Before this Congress was opened, the International Union of Mathematics was founded in New York. German mathematicians, represented by the Deutsche Mathematiker-Vereinigung (DMV), were invited to participate in its inauguration. "We will be able to perform in America completely on an equal footing with the international family of mathematicians," the chairman of the DMV informed the German participants of the Cambridge congress. However, he alerted his colleagues that the atmosphere would not be entirely relaxed: "We cannot expect that we will be greeted friendly by every participant."[2] Heisenberg used this stage to review the progress made in stability theories since his doctoral dissertation—which subsequently appeared in English translation as a NACA-Technical Memorandum (Heisenberg, 1950, 1951). However, the rise of McCarthyism made the scientific relations of American scientists with colleagues from other countries difficult. In 1954 the chairman of the Federation of American Scientists reported that foreign colleagues who were invited to America were "deeply disappointed by the increasing narrowmindedness of U.S. authorities and by the raising of a new paper curtain between America and the rest of the world" (Weisskopf, 1954, p. 68). But this did not impede the rise of the USA in the 1950s to become the world's leading scientific power.

11.3 The Division of Fluid Dynamics of the American Physical Society

In the early decades of the twentieth century, with the emergence of "modern" research fields like relativity, atomic and nuclear physics, classical subjects like hydrodynamics lost ground among physicists. Hydraulics, aerodynamics, naval

---

[2] Kamke (1890–1961) to Heisenberg, 19 July 1950. MPGA, Heisenberg Papers, Correspondence Folder 1950.

architecture, gas dynamics, meteorology—to name only specialities in which turbulence was perceived as a particular challenge—were beyond the scope of physics. The Second World War was called "A Physicists' War" in view of the development of radar and the atomic bomb (Kevles, 1978, Chap. 20). But research on turbulence, shock waves and other investigations from one or another branch of fluid dynamics were also part and parcel of the scientific wartime effort. Physicists realised that they had ignored these specialities on the fringes of their discipline. This neglect motivated representatives of the American Physical Society (APS) to appropriate fluid dynamics as a sub-discipline of physics in a new Division of Fluid Dynamics (DFD). Its foundation illustrates how research areas that were previously viewed mainly from the perspective of engineering have been transferred to the field of physics. Fluid dynamics in general, and turbulence in particular, benefited from this trend. The increasing involvement of physicists in post-war research on turbulence suggests a closer look at the appropriation of fluid dynamics by the community of physicists in the US.

The history of the DFD began in June 1946, when APS president Edward Condon (1902–1974) appointed Kármán, Dryden, Howard Emmons (1912–1998), John von Neumann (1903–1957) and Raymond Seeger (1906–1992) as members of a new Committee on Fluid Dynamics. Each one of them had a record of pertinent war work. Seeger, head of the Aeroballistic Research Department at the Naval Ordnance Laboratory (NOL) at White Oak, Maryland, was chosen as chairman. In August 1946, after returning from the Bikini Atoll, where he was involved in atomic bomb tests, Seeger organized as a first step at the forthcoming annual physics conference of the APS in January 1947 in New York a symposium on "Recent Trends in Fluid Dynamics" with Kármán and Hans Bethe (1906–2005) as invited lecturers. He suggested furthermore "to consider the means of continuing the advancement and diffusion of knowledge on Fluid Dynamics" at the APS business meeting during this conference.[3] In a letter to Bethe, Seeger explained the goals of his Committee:

> Many of us feel that Fluid Dynamics is a subject of considerable physical interest which has been neglected by physicists in the past and revived only during the war. We are eager, therefore, to continue this interest, perhaps by having a permanent committee or even a division of the American Physical Society.[4]

The New York meeting of the APS in January 1947 brought to the fore that fluid dynamics deserved indeed more attention from the physics community. Dryden's collaborators, for example, informed the American physicists at this meeting about their wartime investigation of boundary layer instability. The forthcoming Washington meeting of the APS in May 1947 provided another occasion to demonstrate that there was a continuous and growing interest in fluid dynamics. Seeger got in contact with the editor of the *Journal of Applied Physics* and announced a growing number of papers on fluid dynamics which were likely to be submitted to this journal in the

---

[3] Seeger to Darrow (1891–1982), 28 August 1946. American Physical Society Division of Fluid Dynamics Archives, SC MS 0002, Special Collections, Lehigh University, Bethlehem, Pennsylvania, USA (henceforth abbreviated as DFDA), 4-1.

[4] Seeger to Bethe, 28 August 1946. DFDA, 4-2.

future. "In general," he explained the scope of these papers, "the physics of fluid dynamics is understood to include aerodynamics, hydrodynamics, ballistics, water entry, and explosion phenomena."[5] The journal editor was pleased to learn "that the Committee on Fluid Dynamics is considering the Journal of Applied Physics as its official outlet."[6]

Shortly after the Washington meeting Seeger officially proposed to the Council of the APS the establishment of a Division of Fluid Dynamics. The Council signalled green light at a meeting in Montreal on 20 June 1947. Subsequently, by-laws had to be formulated. The recently established Division of Solid State Physics of the APS served as a role model. In August 1947, Seeger submitted a first draft of by-laws and a circular to the secretary of the APS.[7] In October 1947, the secretary of the APS announced that the DFD has been authorized by the Council; he described its object as "the advancement and diffusion of knowledge of the physics of fluids including aerodynamics and hydrodynamics, phenomena of rarefied gases and explosions, plasticity, et al." and opened the new division for enrollment.[8] The Division's major activity was to organize meetings at least once a year "at such time and place as shall be ordered by the Executive Committee." The meetings should be organized together with the annual APS conferences. At each meeting, a Programme Committee was in charge of the organisation. Official announcements should be published in the *Journal of Applied Physics*, designated for the time being as the official organ of the DFD.[9] By January 1948, the by-laws were approved.[10]

With the organizational problems settled, the focus shifted to topics suitable for sessions at forthcoming Division meetings. Among other subjects, turbulence was chosen as a topic for the a joint meeting with the Aeronautical Society at the forthcoming APS conference, held in January 1949 in New York. As it turned out, the interest in turbulence was so strong that it became split in a two-session symposium on two days during the week of the New York meeting.[11]

The turbulence symposium organized by the DFD at the APS meeting in New York signalled the rise of turbulence as a particular concern for physicists in the USA. A major actor in this effort was Kármán's former student, Francis Clauser (1913–2013), whom the DFD charged with the organization of the New York symposium. Clauser's career started at Johns Hopkins University where he set up a new Department of Aeronautics after the war. Next to supersonics, he regarded turbulence as a most important field of research. In order to firmly establish turbulence on the agenda of his Department, he hired in 1947 as assistant professor Stanley Corrsin (1920–1986),

[5] Seeger to Hutchisson, 3 April 1947. DFDA, 4-2.

[6] Hutchisson to Seeger, 15 April 1947. DFDA, 4-2.

[7] Correspondence between Seeger and Darrow, June to August 1947. DFDA, 4-2 and 4-3.

[8] Circular on the "Formation of a Division of Fluid Dynamics in the American Physical Society," 6 October 1947. DFDA, 4-3.

[9] By-laws, undated (probably 1947). DFDA, 2-8.

[10] Seeger to Darrow, 14 January 1948. DFDA, 4-4.

[11] Emmons to the members of the Executive Committee, 17 August 1948 and 27 December 1948, DFDA, 4-5 and 4-6.

who had encountered turbulence as a research subject in his war work under Kármán at GALCIT (see Sect. 10.1).[12] Clauser was also very eager to make his institute part of the network of military research establishments in his vicinity, such as Seeger's Aeroballistic Department at the NOL with whom he established a "joint venture" in turbulence research. "I believe that some kind of mutual arrangement could and should be worked out," Seeger agreed with Clauser's suggestion. "In that connection, I still have hopes that we will be able to avail ourselves of the part-time services of Corrsin. It turns out that we are becoming interested now with turbulence."[13]

With regard to the forthcoming turbulence symposium in New York, Clauser began in autumn 1948 to address suitable speakers. "I have received acceptances from Chandrasekhar, Liepmann and Kovasznay to give papers," Clauser informed the Chairman of the DFD by November.[14] Liepmann was an old acquaintance of Clauser from common years at GALCIT. Leslie S. G. Kovasznay (1918–1982) was a recent import at Johns Hopkins University. Clauser had just hired him at his Department "to handle a research project on the measurement of turbulence in high speed air streams by means of a hot wire anemometer," he explained in the yearly report of his Department. "Dr. Kovasznay is a graduate of the Royal Hungarian University at Budapest where he was a member of the faculty. Afterwards he was engaged in research work for one year at Cambridge, England, before coming to this country."[15] The Indian astrophysicist Subrahmanyan Chandrasekhar (1910–1995) had come to America already before the war to work at the Yerkes Observatory in Williams Bay, Wisconsin, run by the University of Chicago (Wali, 1991, Chap. 9). Another invitee to the New York turbulence symposium was the Princeton astrophysicist Martin Schwarzschild (1912–1997). Both, Schwarzschild and Chandrasekhar, had just begun to correspond extensively about turbulence as a subject of major concern for astrophysics.[16]

Besides astrophysicists and aeronautical scientists, whose affiliation with turbulence was taken for granted, Clauser also invited the mathematician Mark Kac (1914–1984) and the physicist George Uhlenbeck (1900–1988) to the New York turbulence conference although they had little prior expert knowledge. Kac responded that he knew "next to nothing about turbulence" but nevertheless accepted to deliver an invited lecture because Clauser regarded Kac's speciality of random noise as most

[12] Clauser to Kouwenhoven (1886–1975), 26 May 1947. Ferdinand Hamburger Archives, Sheridan Libraries, Johns Hopkins University. Record Group 06.080, Department of Aeronautics, Subgroup 2, Series 1 (Professional Papers of Francis H. Clauser, 1946–1961, henceforth abbreviated as Clauser Papers), Folder: Kouwenhoven, William B., 1947–1953. On Corrsin's career and work on turbulence see Meneveau and Riley (2011).

[13] Seeger to Clauser, 26 September 1947. Clauser Papers, Folder: National Science Foundation, 1952–1961.

[14] Clauser to Emmons, 2 November 1948. Clauser Papers, Folder: Turbulence Symposium, 1948–1949.

[15] Clauser Papers, Folder: Isaiah Bowman, Report to, 1948–1949.

[16] This correspondence is preserved in the Special Collections Research Center, University of Chicago Library, Subrahmanyan Chandrasekhar Papers 1928–1995 (henceforth abbreviated as Chandrasekhar Papers), Box 28, Folder 13.

pertinent.[17] Uhlenbeck was invited for his expert knowledge in statistical mechanics: "At present there is an effort to carry over into turbulence many of the concepts that have been developed in statistical mechanics," Clauser explained, "I believe that many of the people interested in turbulence would be eager to hear a discussion of the conceptual difficulties that now exist in the statistical mechanics and that would continue to be troublesome when application is made to turbulence."[18]

Although Chandrasekhar and Uhlenbeck finally could not attend, the invitations reveal the intention to reach out for physicists with quite different orientations. The lecturers at the New York turbulence symposium were Corrsin ("Some measurements in a round turbulent jet"), Kac ("Statistical analysis of random functions"), Kármán and Lin ("Statistical theory of isotropic turbulence"), Kovasznay ("Optical methods of measuring turbulence"), Schwarzschild ("Turbulence in the atmosphere of stars") and G. C. Williams ("Combustion-generated turbulence in relation to flame propagation") (APS, 1949, p. 1282).

The DFD regarded the meeting so successful that turbulence entered the agenda of future conferences as a subject of special sessions. In summer 1949, for example, at the inauguration of new facilities of Seeger's Department on Aeroballistics, the DFD convened a two-day conference at the NOL with another special session on turbulence chaired by Burgers. The invited speakers at this event were Kampé de Fériet ("Spectral Tensor of a Homogeneous Turbulence"), Batchelor ("The Nature of Turbulent Motion at Large Wave-Numbers"), Chandrasekhar ("Development of Heisenberg's Theory of the Decay of Isotropic Turbulence"), Squire ("Investigation of the Turbulence Characteristics of an Experimental Low-Turbulence Wind Tunnel") and Frenkiel ("Some Remarks on Turbulent Diffusion").[19]

In June 1949, the American Physical Society held its Semi-Centennial Meeting in Cambridge, Massachusetts. Dryden, in the capacity as the acting chairman of the Division of Fluid Dynamics of this year, used the opportunity for a review: He praised Seeger, "a physicist who had become interested in fluid dynamics as a result of his wartime work," as the "leading spirit" behind the formation of the DFD and hoped that it resulted in a "permanent consideration of fluid dynamics by physicists." He discerned turbulence as a problem which deserved "greater attention from the most capable physicists" and pointed to Heisenberg as role model of a physicist whose reputation in quantum physics did not prevent him from pioneering work in the classical field of turbulence. Alluding to Heisenberg's internment at Farm-Hall he mused that Heisenberg was "fortunately for turbulence research forcibly diverted from nuclear research and compelled to seek new interests." He concluded with a kind of teaser for physicists to engage in turbulence research:

> I have not time to review the full picture developed by Heisenberg, Onsager, and others, of the concept of a spectrum of energy fed at the larger wave length and by the separation of flow and regular eddying motions determined by the geometry of the apparatus, the energy passes

[17] Kac to Clauser, 19 November 1948. Clauser Papers, Folder: Turbulence Symposium, 1948–1949.

[18] Clauser to Uhlenbeck, 16 November 1948. Clauser Papers, Folder: Turbulence Symposium 1948–49.

[19] DFDA, 1-3 and 4-7.

through a cascade of eddies of smaller and smaller wave length as a quasi-stationary process, until at the short wave length and the dissipation action of viscosity makes its influence felt sufficiently to rapidly reduce the energy at high frequencies. With certain assumptions, this theory can be given quantitative form and subjected to experimental check. [...]

There are many more puzzling questions in turbulent fluid motion, in particular, the exact source of the randomness and the nature of the statistics which describes the energy distribution in the different modes. I think I have said enough to indicate that physicists can find challenging problems and can make significant contributions.[20]

[20] Draft of speech, 17 June 1949. Dryden Papers, Special Collections, The Milton S. Eisenhower Library, The Johns Hopkins University, Pubs-files, 1949, box 11, folder "Div of Fluid Mechanics—Amer. Physical Soc.—Cambridge 1949".

Chapter 12
Turbulence in the 1950s

Abstract By the 1950s turbulence was considered ripe for systematic presentation in monographs. The *Journal of Fluid Mechanics* and *The Physics of Fluids* offered new outlets for publishing research on turbulence. At joint symposia of IUTAM with the International Astronomical Union (IAU) and the International Union of Geodesy and Geophysics (IUGG) turbulence transgressed disciplinary boundaries.

12.1 The First Textbooks on Turbulence

Batchelor used the inauguration conference at the Naval Ordnance Laboratory in summer 1949 to promote K41 in the USA. "You will have seen previously from the program of the dedication ceremonies that Dr. Batchelor is speaking in the turbulence symposium, on the Cambridge work in extending Kolmogoroff's theory," Ronald Smelt (1913–2005), deputy chief for aeroballistics research at the NOL, wrote to Clauser. "He has asked me to arrange, if possible, for some lectures during his stay in the USA."[1]

Batchelor's promotion of Kolmogorov's theory can hardly be exaggerated as an effort to stage turbulence as a challenge for further investigation. "When you are in Cambridge, you should try to see Batchelor, he is an awfully good man and very sound contrary to the other Cambridge luminaries in astrophysics," Chandrasekhar advised Schwarzschild in May 1950 when Schwarzschild went on a trip to Europe.[2] When Taylor asked Chandrasekhar in September 1951 for an evaluation of Batchelor's application for a fellowship at Trinity College, Chandrasekhar was again full of praise: "I am very well acquainted with Dr. Batchelor's work on the theory of turbulence. Indeed, when about two years ago I started seminar classes on 'The theory of turbulence', I found that the most satisfactory way to go about learning the subject was to read and discuss the various papers of Dr. Batchelor, one by one."[3]

---

[1] Smelt to Clauser, 13 April 1949. Clauser Papers, Folder: Smelt, Ronald, 1949.

[2] Chandrasekhar to Schwarzschild, 1 May 1950. Chandrasekhar Papers, Box 28, Folder 12.

[3] Chandrasekhar to Taylor, 24 September 1951. Chandrasekhar Papers, Box 31, Folder 3.

M. Eckert, *Turbulence—an Odyssey*, History of Physics,
https://doi.org/10.1007/978-3-030-91459-2_12

By the time of this praise Batchelor's list of publications comprised fourteen papers, all of them on turbulence; four papers were co-authored with Albert Alan Townsend (1917–2010), one with Robert William Stewart (1923–2005), both advanced students at Cambridge, the remaining nine by himself as a single author. In 1950 he was awarded the prestigious Adams Prize of the University of Cambridge for an essay that became the first draft of *The Theory of Homogenous Turbulence* (Batchelor, 1953). Shortly after the revelation presented at the Paris mechanics conference in 1946, Batchelor had reviewed Kolmogorov's theory in the *Proceedings of the Cambridge Philosophical Society* (Batchelor, 1947)—this was the paper that disseminated the details of K41 to the Western world. Even in the USSR turbulence researchers used a Russian translation of Batchelor's paper because it made the theory more understandable than Kolmogorov's original papers (Moffatt, 2002). In the following years the theory became subject of close scrutiny. By the early 1950s it was ripe for a comprehensive review. In the preface of *The Theory of Homogenous Turbulence*, written in July 1952, Batchelor regarded this "an opportune time to put some of the existing work into more permanent form", not the least for these reasons:

> It has been my experience that students and others coming fresh to the subject have found a need for a systematic and complete account of the established knowledge of turbulence, and I hope that this book will fill part of that need. There is perhaps another reason why the present time is suitable for the appearance of a book on homogeneous turbulence. It seems that the surge of progress which began immediately after the war has now largely spent itself, and there are signs of a temporary dearth of new ideas. (Batchelor, 1953, p. ix)

The Theory of Homogenous Turbulence became a textbook classic (Fig. 12.1). However, Chandrasekhar's admiration of Batchelor waned. He became "disillusioned very rapidly by the sterility of his approach to scientific problems."[4] But there may have been some rivalry involved because Chandrasekhar planned himself a textbook on the statistical theory of turbulence. It was doomed when Batchelor's monograph appeared in 1953. "I have since changed my mind and would rather write my book on stability first," Chandrasekhar informed the publisher.[5] Chandrasekhar's *Hydrodynamic and Hydromagnetic Stability* appeared in 1961 and also became a classic (Chandrasekhar, 1961). His plan for a book on the statistical theory of turbulence never materialized.

Shortly after Batchelor, his Cambridge fellow student and frequent co-author, Townsend, published a monograph on *The Structure of Turbulent Shear Flow* (Townsend, 1956). It built on the knowledge of small-scale turbulence as described in Batchelor's book, but extended it so that turbulent shear flow at larger scales as observed in experiments came into focus. "This monograph will undoubtedly stand as a cornerstone for much of the serious research on shear flows in the years to come", Clauser praised it in *Nature* (Clauser, 1957).

Another textbook on turbulence originated from a series of lectures addressed to chemical engineers of the Dutch Shell laboratories at Amsterdam and Delft in 1950. The lecturer was Hinze, a former student of Burgers. In 1952 Hinze became

[4] Chandrasekhar to O. J. Eggen, 13 September 1968. Chandrasekhar Papers, Box 11, Folder 11.
[5] Chandrasekhar to Clarendon Press, 10 December 1956. Chandrasekhar Papers, Box 68, Folder 8.

Fig. 12.1 George Batchelor's monograph became a textbook classic. It is still reprinted and used as an introduction to K41

director of a new engineering department at the Shell laboratories where he focused both on the foundations of turbulence and its applications. In 1956 he returned to the technical university in Delft as professor of fluid mechanics (Alkemade, 2019, p. 45). In his lecture course at Shell he attempted to provide the engineers with "a sufficient theoretical basis not only for studying the specialized literature on turbulence but also for theoretical investigations on those chemical engineering problems in which turbulence plays an essential part, such as mixing and heat and mass transfer" (Hinze, 1959, p. v).

Batchelor's, Townsend's and Hinze's textbooks concerned the statistical theory of fully developed turbulence. Transition to turbulence became a research field in its own right which was regarded ripe for of a textbook representation by the same time. Hinze even hesitated to attribute to his monograph the character as a textbook because "instability theories and transition from laminar to turbulent flow have not been considered" (Hinze, 1959, p. v). The missionary of the latter was Lin. His role for the dissemination of stability theory parallels that of Batchelor for the statistical theory of turbulence. Like Batchelor, whose authoritative representation was envied by Chandrasekhar, Lin's role was challenged by a rival of considerable reputation: Tollmien. In summer 1947, Tollmien was appointed as Prandtl's successor to the chair of applied mechanics at Göttingen University. "You must feel proud that your early theoretical work has been so fully confirmed", Dryden congratulated Tollmien with a hint to the wartime experiments of Schubauer and Skramstad.[6] "I had never hoped to see such a satisfactory confirmation of the stability theory," Tollmien responded.[7] But he was far from enthusiastic about Lin's review of stability theory published in subsequent issues of the *Quarterly of Applied Mathematics* (Lin, 1945, 1946).

When Lin heard rumors about Tollmien's objections and asked for the reasons, Tollmien detailed in a five-page letter, written in English, what upset him. In particular, he criticized Lin's version of the historical development. "The history of the stability research of laminar flows is rather complicated," he began his critique. Lin had attributed the priority to Heisenberg, but Tollmien insisted that Prandtl and Tietjens, not Heisenberg, deserved the primary credit. "Heisenberg was, during the early twenties, very often at Göttingen and was completely familiar with the ideas current at this time in the Prandtl school on the stability problem." Tollmien further criticized that Lin had not mentioned Noether's "frightening criticism" which dealt an "almost deadly blow" to the whole approach. He also distanced his own work from Heisenberg's "doubtful expansions" and claimed his own priority for having elaborated the crucial physical concept of the "inner friction layer" which resulted in the first calculation of the neutral stability curve. "Summarizing the whole situation, I think it is really silly to suppress my name in the history of the stability problem in favour of Heisenberg's or yours as it has been done in several American publications."[8] Lin attempted to justify Heisenberg's primacy but finally gave in: "Since you and others think that Heisenberg's expansion method and your method are different, I must admit that you did not 'adopt' a well established method. I must apologize to you for using that word."[9]

Beyond the dispute about approximation methods in stability theory, the controversy exposed other concerns. "In 1945 many foreigners believed that German science was dead and that one can get away with plundering in our garden," Tollmien wrote to a colleague.[10] The old rivalry between Prandtl and Kármán may also have

[6] Dryden to Tollmien, 27 August 1947. AMPG, Abt. III, Rep. 76B, box 1 (henceforth abbreviated as Tollmien papers).

[7] Tollmien to Dryden, 9 April 1948. Tollmien papers.

[8] Tollmien to Lin, 24 November 1948. Tollmien papers.

[9] Lin to Tollmien, 7 June 1949. Tollmien papers.

[10] "1945 glaubten eben viele Ausländer, dass die deutsche Wissenschaft tot wäre und dass man ungestraft in unserem Garten plündern könnte." Tollmien to Willers, 4 April 1949. Tollmien papers.

played a role. "I think you were ill advised", Tollmien remarked in one letter, blaming Kármán as Lin's thesis adviser.[11]

Tollmien presented extensive reviews on stability theory at a meeting of GAMM in 1951 and the IUTAM congress in 1952 (Tollmien, 1952, 1953), but in 1955 Lin retained the primacy over this field with his monograph on *The Theory of Hydrodynamic Stability* (Lin, 1955). "There can be no reasonable doubt about the usefulness of Professor Lin's monograph, for it is the first major survey of the theory of hydrodynamic stability to appear in book form," a reviewer praised it in the *Journal of Fluid Mechanics*. "The work as a whole bears the stamp of authority" (Owen, 1956).

12.2 New Journals

The *Journal of Fluid Mechanics* is, according to its website, "arguably the most highly-regarded and influential journal in the field that is read by scholars world-wide."[12] It was founded in 1956 at the initiative of Batchelor who also served for four decades as its editor. At the time of its foundation it provided a new stage for disseminating news and articles on all aspects of fluid mechanics, not the least on turbulence. "I remember that before 1956 I was disturbed by what seemed to me to be an unnatural and harmful three-way split of the literature on fluid mechanics," Batchelor later recalled what motivated him to found a new journal. The split was between theory, experiment and application:

> There were journals for each group separately, but none which embraced all three except some journals of very wide scope covering much more than fluid mechanics. It was annoying to have to chase after so many different journals in libraries in order to keep up with developments, and when it came to submission of a paper by one of the turbulence group at Cambridge there never seemed to be a journal that was wholly appropriate. (Batchelor, 1981, p. 2)

Other researchers on one or another topic of fluid mechanics shared this experience. When the Division of Fluid Dynamics of the American Physical Society was founded in 1947, it resorted to the *Journal for Applied Physics* as outlet for publications. But this situation was soon regarded as unsatisfactory. According to the minutes of a DFD-meeting in January 1949 there was "a feeling among a considerable number of members of the Fluid Dynamics Division that there was a real need for a new journal in this field."[13] In 1956 the American Institute of Physics (AIP) inquired by a survey how American physicists used the available journals and what

---

[11] Tollmien to Lin, 24 July 1950. Tollmien papers.

[12] https://www.cambridge.org/core/browse-subjects/mathematics/fluid-mechanics-hub/george-batchelor-centenary-event.

[13] G. H. Shortley to the Members of the Committee on Publications in Fluid Dynamics, 4 April 1949. Physics of Fluids Records, 1956–1981, American Institute of Physics, Niels Bohr Library & Archives, College Park, Maryland (henceforth abbreviated: Physics of Fluids Records). Folder: Correspondence prior to January 2, 1958.

needs they expressed for the future. As a result the AIP envisioned "a new pattern of journal publications."[14] The DFD seized the opportunity and revealed their plan of a "Journal of Fluid Physics" with a broad scope. "We would wish to see included in the proposed journal basic research papers in magnetohydrodynamics, shock wave phenomena, compressible and high temperature fluid flows, plastic flow, turbulence, liquid state physics, ionized fluid and plasma flows, as well as certain basic aspects of fluid physics bordering geophysics and astrophysics."[15]

Frenkiel played in this initiative a similar role like Batchelor in England. In the capacity as Secretary of the DFD Frenkiel became the major actor in the USA for the foundation of the planned journal. Like Batchelor he had actively contributed to turbulence research and presented papers at preceding IUTAM Congresses and numerous other occasions. His career had started under tragic circumstances in the Second World War (Emrich et al., 1987). After his doctoral work with Kampé de Fériet and the German invasion of France in 1940 he had moved with his mentor to the Aeronautical Research Station at Toulouse. When the German occupation was extended to Southern France, he was imprisoned and spent the remaining two years of the war in concentration camps. His wife and their unborn child were killed by the Nazis. After the war, Frenkiel emigrated to the United States where he was employed for short periods at the Department of Aeronautical Engineering of the Cornell University and at the Naval Ordnance Laboratory. In 1950, he joined the Applied Physics Laboratory of the Johns Hopkins University at Silver Spring which had close ties with Clauser's Department of Aeronautics. The latter move was once more caused by the political events in the Cold War: In the course of Project "Paperclip" the NOL employed German aerodynamicists who had participated in the V-2 project at Peenemünde. One of them, Hermann H. Kurzweg (1908–2000) became Frenkiel's chief—a situation which he was unwilling to tolerate. "And five years have not wiped out from my memory Buchenwald and Auschwitz," Frenkiel wrote to Kármán shortly before he left the NOL.[16]

As Secretary of the DFD, Frenkiel made the plan for the new journal his mission. Its scope should cover a broad spectrum of research fields, including turbulence, with the focus on the underlying physics. "Emphasis should not be on applied mathematics but on that combination of experiment, conceptual models, and formal analysis that is thought of as characterizing good physics."[17] Batchelor anticipated an unpleasant rivalry with his *Journal of Fluid Mechanics*. As Frenkiel was alerted by a colleague, "Dr. Batchelor took the view that it might be wise to wait and see whether the Journal of Fluid Mechanics could meet the needs in this field before setting up

[14] Memo by J. H. McMillen, 20 March 1956. Physics of Fluids Records, Folder: Correspondence prior to January 2, 1958.

[15] Frenkiel (Secretary of the DFD) to Frederick Seitz (1911–2008) (Chairman of the AIP's Governing Board), 27 April 1956. Physics of Fluids Records, Folder: Correspondence prior to January 2, 1958.

[16] Frenkiel to Kármán, 17 February 1950. TKC, 94-12.

[17] Proposal for a Journal of Fluid Physics, 10 May 1956. Physics of Fluids Records, Folder: Correspondence prior to January 2, 1958.

another journal."[18] Batchelor also informed Frenkiel directly about the background and tendencies of his journal.[19]

But the "wait-and-see" attitude did not fit well with the AIP intentions which concerned the need for new journals also in other physical sub-disciplines. Nor did it please Frenkiel who circulated a questionnaire to the members of the DFD concerning the scope and title of the new journal. Batchelor was "dismayed and disturbed" about this move and appealed to Frenkiel as a representative of their common field of research. "Do you not think that the Fluid Dynamics Division would be doing science a service by strengthening an international Journal rather than by setting up another national journal?"[20] But Frenkiel was already carrying the initiative a step further. He formed a committee with the task to present the Governing Board of the AIP with a more definite proposal about the new journal. At the first meeting of this committee in February 1957, Batchelor's concerns were dismissed. There was "unanimous agreement" that the proposed AIP journal and the *Journal of Fluid Mechanics* could both flourish.[21] Three weeks later, the Governing Board of the AIP authorized the establishment of the new journal, titled "The Physics of Fluids"; the first issue was due to appear in January 1958.[22]

Batchelor was frustrated. "In these circumstances," he wrote to Frenkiel, "it is perhaps not necessary to go on talking about the pros and cons of the new journal." In an effort to cut his losses he proposed a division of labour. The *Journal of Fluid Mechanics* would focus "on the mechanical aspects of fluid motion" and the new AIP-Journal "on the more physical aspects." Although he admitted that such a discrimination was problematic, he hoped that this would reserve most of the articles on fluid mechanics proper, such as "hydrodynamics, dynamics of compressible fluids, boundary layer and turbulent phenomena," for the *Journal of Fluid Mechanics*.[23] If such a division of labor had been adopted, Batchelor's *Journal of Fluid Mechanics* would have become the main Anglo-American publication outlet for turbulence research. However, Frenkiel responded that he had "no authority to commit the future editorial policy of the journal" and expressed his confidence that the competition between both journals would be "of a constructive nature."[24] The

---

[18] Schubauer to Frenkiel, 26 September 1956. Physics of Fluids Records, Folder: Correspondence prior to January 2, 1958.

[19] Batchelor to Frenkiel, 4 December 1956. Physics of Fluids Records, Folder: Correspondence prior to January 2, 1958.

[20] Batchelor to Frenkiel, 7 February 1957. Physics of Fluids Records, Folder: Correspondence prior to January 2, 1958.

[21] Minutes of the Meeting of the AIP Planning Committee, 21 February 1957. Physics of Fluids Records, Folder: Correspondence prior to January 2, 1958.

[22] Frenkiel to Uhlenbeck and others, 5 April 1957. Physics of Fluids Records, Folder: Correspondence prior to January 2, 1958.

[23] Batchelor to Frenkiel, 3 May 1957. Physics of Fluids Records, Folder: Correspondence prior to January 2, 1958.

[24] Frenkiel to Batchelor, 13 May 1957. Physics of Fluids Records, Folder: Correspondence prior to January 2, 1958.

next day he drafted a memo about the editorial organization and scope of *The Physics of Fluids*:

> The scope of these fields of physics includes: hydrodynamics, dynamics of compressible fluids, shock and detonation wave phenomena, hypersonic physics, rarefied gases and upper atmosphere phenomena, transport phenomena, hydromagnetics, ionized fluid and plasma physics, liquid state physics, superfluidity, boundary layer and turbulence phenomena, as well as certain basic aspects of physics of fluids bordering geophysics, astrophysics, biophysics and other fields of science. Emphasis will not be on applied mathematics, but on that combination of experiment, conceptual models, and formal analysis that is thought of as characterizing good physics.[25]

On 1 July 1957, Frenkiel announced the foundation of *The Physics of Fluids*.[26] Disregarding the rivalry between Batchelor and Frenkiel as editors of the two journals, turbulence ranked high on the respective list of topics. In the four years from 1958 to 1961 both journals published 76 articles concerned with turbulence (35 in *The Physics of Fluids*, 41 in the *Journal of Fluid Mechanics*). As Frenkiel had predicted, there was no dearth of articles. Both journals flourished and had to enlarge their volume in the following years and decades (Scott, 2008; Davis and Pedley, 2006).

12.3 Transgressing Disciplinary Boundaries

Despite the surge of turbulence research in the 1950s, turbulence was not yet a discipline in its own right. By their institutional affiliation, turbulence researchers belonged to aeronautical engineering, applied mathematics, physics or some other discipline. As illustrated by the formation of the Division of Fluid Dynamics, the American Physical Society appropriated turbulence as part of fluid dynamics only after the Second World War under the umbrella of physics. Turbulence came of age by transgressing disciplinary boundaries. Transdisciplinarity was (and still is) a characteristic feature of fluid dynamics at large, and turbulence in particular.

This is illustrated by the joint symposia that IUTAM (see Sect. 11.2) co-organized with other International Unions. A first joint symposium that IUTAM co-organized with the International Astronomical Union (IAU) was held in August 1949 in Paris on "Problems of Cosmical Aerodynamics". It was meant to explore "the still undeveloped nature of the border land between astronomy and gas dynamics". The organizers of this event, Burgers and the Dutch astronomer Hendrik Christoffel van de Hulst (1918–2000), emphasized how important the "present-day developments in hydro- and aerodynamics" were for astronomers. "In particular, the problems of turbulence and those of expansion phenomena and of shock waves immediately come to the foreground" (Burgers and van de Hulst, 1949, p. i). At Burgers's request von Kármán presented an introductory lecture "on possible applications of the theory of turbulence, as developed for the understanding of fluid motions under terrestrial

---

[25] Draft, 14 May 1957. Physics of Fluids Records, Folder: Memorabilia for The Physics of Fluids.
[26] Announcement, 1 July 1957. Physics of Fluids Records, Folder: Board of Editors.

conditions, to celestial problems" (Burgers and van de Hulst, 1949, p. 129). Kármán turned it into a survey on the recent progress in the statistical theory of turbulence—which made it difficult for Batchelor as the following speaker to stick to his original outline: "It is not a very enviable task to follow Dr. von Kármán on this subject of turbulence," Batchelor began his lecture, because Kármán covered "so many matters that the list of things which I had to say is now torn to shreds by the crossings out I have had to make as his talk progressed" (Burgers and van de Hulst, 1949, p. 149). So Batchelor dedicated the major part of his talk to a novel research field for astronomers as well as turbulence researchers, the interaction between magnetic fields and turbulence. Another remarkable lecture was presented by von Weizsäcker on "Turbulence in Interstellar Matter" (Burgers and van de Hulst, 1949, pp. 158–164, 200–203).

The encounter of turbulence research with astronomy was regarded so fruitful that it became the beginning of a series of symposia on "Cosmical Gas Dynamics". The following joint IUTAM-IAU-Symposia were held in Cambridge, UK (1953), Cambridge, Massachusetts, USA (1957), Varenna, Italy (1960), Nice, France (1965) and at the Crimean Astrophysical Observatory, USSR (1969). Besides astronomy there were other disciplines with a genuine interest in fluid dynamics, in particular meteorology and geophysical disciplines. Next to the IAU, the International Union of Geodesy and Geophysics (IUGG) became a regular partner of IUTAM for joint events. They co-organized a Colloquium on "Plastic Flow and Deformation within the Earth" in Hershey, Pennsylvania, USA (1950), an International Symposium on "Atmospheric Diffusion and Air Pollution" in Oxford, UK (1958) and on "Fluid Mechanics in the Ionosphere" in Ithaca, N.Y., USA (1959).[27] With the proclamation of 1957 as the International Geophysical Year (IGY), fluid dynamics moved closer to geophysics. "It may be significant that the active operation of our Editorial Board happens to start on the first day of the International Geophysical Year," Frenkiel remarked in his editorial announcement about his journal on the same day. "The scope of The Physics of Fluids includes several fields of fundamental importance to theoretical and experimental geophysics and it should be expected that our journal will contribute in some measure to the scientific interpretation and to the understanding of the numerous data obtained during IGY."[28]

Hence it is not accidental that turbulence, too, rose to prominence in a joint venture with geophysics. The sessions at the IUTAM-IUGG-symposium on "Atmospheric Diffusion and Air Pollution" in 1958 and the names of the respective chairmen reflect the broad concern about turbulent diffusion (Frenkiel and Sheppard, 1959, p. xiii):

General survey of atmospheric diffusion and pollution (G. I. Taylor)

Recent findings on atmospheric turbulence (J. Kampé de Fériet)

Theory of turbulent diffusion (G. K. Batchelor)

Diffusion of heavy or finite particles (A. M. Obukhov)

Transfer through the troposphere and stratosphere (H. R. Byers, C. H. B. Priestley)

---

[27] https://iutam.org/publications/proceedings-of-iutam-symposia/.

[28] Announcement, 1 July 1957. Physics of Fluids Records, Folder: Board of Editors.

Effects of thermal stratification on diffusion (K. Wieghardt)

Pollution patterns from point and area sources (H. Solberg)

Review of previous sessions (G. Sutton)

The subsequent symposium on "Fluid Mechanics in the Ionosphere" in 1959 also expressed a great interest in turbulence. Batchelor introduced the talks on fluid mechanics with the remark that "turbulence is bound to be the most important single topic, so that the objective of the six fluid mechanics speaker is to give a short course on turbulence—turbulence, of course, in a broad sense, allowing for the effects of the earth's magnetic field, the density gradient, and several other factors as well."[29] The papers presented at the symposium were published in the *Journal of Geophysical Research*. Almost half of them dealt with turbulence.

In summer 1960 the General Assembly of IUGG and IUTAM, respectively, approved preliminary plans for a joint symposium on turbulence and committed a scientific committee, chaired by Frenkiel, with the organisation. By the same time Favre, professor of fluid mechanics, planned to convene the crème de la crème of turbulence research to an international colloquium for the inauguration of his new Institut de Mécanique Statistique de la Turbulence (IMST) at Aix-Marseille University. The organizers of both events therefore decided to choose Marseille as the common site for these conferences. Favre's International Turbulence Colloquium was scheduled for the last week of August 1961, the IUGG-IUTAM Symposium on Fundamental Problems in Turbulence and their Relation with Geophysics for the first week of September 1961 (Frenkiel, 1962). Thus Marseille became the stage for a unique presentation of turbulence.

[29] Transactions of the International Symposium on Fluid Mechanics in the Ionosphere. In: Journal of Geophysical Research, 64:12, 1959, p. 2044.

Chapter 13
Marseille ’61

Abstract In summer 1961 Marseille became the stage for a remarkable two-part event in the history of turbulence: the International Turbulence Colloquium and the Symposium on Fundamental Problems in Turbulence and their Relation with Geophysics, the latter co-organised by the international unions of geophysics (IUGG) and mechanics (IUTAM). It provided an opportunity to review past achievements and discern the problems for future research.

13.1 Turbulence—“A New Science”

In 1961 the Cold War was in full swing. Relations between the USSR and the USA severely deteriorated when Nikita Khrushchev (1894–1971) and John F. Kennedy (1917–1963) ended a summit in June 1961 in a hostile atmosphere. Two months later the Berlin Crisis climaxed when the German Democratic Republic erected a wall between East and West Berlin.

In this atmosphere even a scholarly meeting attended by scientists from the hostile superpowers like that in Marseille on turbulence aroused public curiosity. “At the Faculty of Sciences in Marseille from 28 August to 2 September: The most eminent scientists from 32 countries will participate in a symposium on the fluid mechanics of turbulence,” a newspaper announced this “gathering at the highest level”.[1] The Russian scientists attracted particular attention. At the opening session “some absences were noticed,” in particular that of Kolmogorov. His former doctoral student, Mikhail Dmitrievich Millionshchikov, gave the journalist the following explanation:

> Professor Kolmogorov cannot travel by plane, so he had to come by train and cross the Federal Republic of Germany. However, he has still not received his transit visa, which is causing this delay, because we hope that it will only be late. Professor Kolmogorov, who is

[1] “À la Faculté des sciences de Marseille du 28 août au 2e septembre : Les plus éminents savants de 32 pays participeront à un colloque de mécanique des fluides turbulents. [...] Un rassemblement à l’échelon le plus élevé”. Le Méridional, 27 August 1961.

M. Eckert, *Turbulence—an Odyssey*, History of Physics,
https://doi.org/10.1007/978-3-030-91459-2_13

> well known for his work on turbulence, is also a great mathematician. I do not understand such a measure and greatly regret this delay.[2]

With regard to the theme of the conference, turbulence, Millionshchikov added:

> I think that such a form of congress is very useful. It is the first time that a congress is reserved for this discipline alone in this form. I am sure that from this work, from the exchange of views, extremely valuable elements will come out for the scholars of this branch. More and more the importance of turbulence appears as one of the fundamental parts of modern physics: Astronautics, aviation, atomic industry need our studies. This shows the importance of such a symposium.[3]

Neither the choice of Marseille as the site for this event nor Favre's commitment as its organizer was accidental. Favre had made the first steps of his career in Marseille at the university's Institute of Fluid Mechanics founded in 1930 at the request of the French Air Ministry. In the 1950s he witnessed the surge of turbulence research particularly in the USA. "We had a very pleasant visit here from Professor Favre from Marseille," wrote Liepmann in July 1954 after a two-months-sojourn of Favre at the GALCIT. "He has made some really nice space-time correlation measurements and his equipment for these measurements seems very good and at present pretty unique. Favre told me that ONERA wants him to go into high speed work on turbulence."[4] Encouraged by ONERA, the French analogue of NASA, Favre planned the establishment of an Institute of Statistical Mechanics of Turbulence (Institut de Mécanique Statistique de la Turbulence, IMST) with a powerful wind tunnel. Both were inaugurated on 28 August 1961 in the presence of the world's elite of turbulence research whom Favre had invited for this event. "Bringing together international leaders," a newspaper titled a report about the inauguration event. "The Colloquium on the mechanics of turbulence opened yesterday. I.M.S.T. starts up its new wind tunnel, the only one of its kind in the world." The International Turbulence Colloquium (28 August–2 September) was only the "prelude" to the International Symposium on Fundamental Problems of Turbulence and their Relation to Geophysics (4–9 September) organised by the International Unions. For the newspaper this double-event signalled a new role for the city. "Marseille must become more and more the city of congresses."[5]

---

[2] "Le Professeur Kolmogorov ne peut pas voyager par avion, il lui fallait donc venir au train et pour cela traverser la République fédérale allemande. Or, il n'a toujours pas reçu son visa de transit ce qui occasionne ce retard, car nous espérons que ce ne sera qu'en retard. En effet le professeur Kolmogorov, très connu pour ses travaux sur la turbulence, est aussi un très grand mathématicien. Je ne comprends pas une telle mesure et déplore grandement ce retard." La Marseillaise, 29 August 1961.

[3] "Je pense qu'une telle forme de congrès et très utile. C'est la première fois qu'un congrès est réservé à cette seule discipline dans cette forme. Je suis sur que de ces travaux, des échanges de pointe de vue, il sortira des éléments extrêmement précieux pour les savants de cette branche. De plus en plus l'importance de la turbulence apparaît comme une des parties fondamentales de la physique moderne : Astronautique, aviation, industrie atomique ont besoin de nos études. C'est dire l'importance d'un tel colloque." La Marseillaise, 29 August 1961.

[4] Liepmann to Kármán, 15 July 1954. TKC, 18–20.

[5] "Groupant les sommités internationales. Le colloque de mécanique de la turbulence s'est ouvert hier. L'I.M.S.T. met en marche sa nouvelle soufflerie d'un type unique au monde [...] Organisé

From a scientific perspective, turbulence was regarded ripe for an inauguration too. The gathering at Marseille served the purpose to stage turbulence as "a new science," to quote Favre with his opening address:

> The Mechanics of Turbulence has developed because of the importance of the scientific problems posed by turbulence in Fluid Mechanics and also in other sciences such as Geophysics, Astrophysics, Magnetohydrodynamics, Thermodynamics, Acoustics, and it has specialised to deal with these problems by using the methods that are relevant in Statistics; it now constitutes a new science.[6]

The Symposium that followed the Colloquium and the inauguration of Favre's institute was opened by Frenkiel in the same vein: "The three events – the Colloquium on the Mechanics of Turbulence, the dedication of the Institute of Statistical Mechanics of Turbulence, and the IUGG-IUTAM Symposium – have great significance to the future progress in turbulence research" (Frenkiel, 1962, p. 3008). With the exception of some papers that were published separately in different journals, the contributions to the IUGG-IUTAM Symposium were published in 1962 in a special issue of the *Journal of Geophysical Research* (volume 67, issue 8, pages 3007–3235); those to the preceding Colloquium were printed under the title *Mécanique de la Turbulence* in a proceedings volume of 468 pages (Favre, 1962). Never before had turbulence become the exclusive subject of such extensive presentation and discussion. Accordingly it is remembered by those who performed on this stage. "I still see the 1961 Marseille meeting as a watershed for research in turbulence," one participant recalled forty years later (Moffatt, 2002, p. 22).

Part of the excitement resulted from the way how the "new science" of turbulence was set in scene. While the IUGG-IUTAM Symposium was focused on geophysical applications, the goal of the Colloquium was more challenging. Preparations had started a year before. The surge of recent research was filtered by the demand that a contribution had to "clarify or extend the fundamental ideas under discussion." Only "highly qualified participants" were invited who represented major research areas. The institutional affiliation of the participants shows that the declared emphasis on "fundamental ideas" did not restrict the invitation to academic institutes; it includes facilities like the Space Technology Labs, the Applied Mathematics Laboratory of the David Taylor Model Basin, the Jet Propulsion Laboratory, the Martin Company, the Royal Aircraft Establishment, the Office National d'Etudes et de Recherches Aéronautiques (ONERA), the Commissariat à l'Energie Atomique, the Aerodynamische

comme prélude a l'inauguration officiel de l' IMST et de sa soufflerie, le colloque dont sept commissions préparent les travaux depuis plus d'un an aura un prolongement du 2 au 9 septembre avec le symposium international sur les problèmes fondamentaux de la turbulence et leur relations avec la géophysique : ce symposium est organisé par les unions internationales de géophysique et de mécanique [...] Marseille se doit d'être de plus en plus la ville des congres." Le Provencal, 29 August 1961.

[6] "La Mécanique de la Turbulence s'est développée en raison de l'importance des problèmes scientifiques posée par la turbulence en Mécanique des Fluides et aussi dans d'autres sciences comme la Géophysique, l'Astrophysique, la Magnétohydrodynamique, la Thermodynamique, l'Acoustique, et spécialisée en utilisant pour traiter ces problèmes des méthodes relevant de la Statistique ; elle constitue maintenant une nouvelle science." (Favre, 1962, p. 5).

Versuchsanstalt and the Deutsche Versuchsanstalt für Luftfahrt. The organizers of the Colloquium also did not qualify the participation along the divide of theory versus experiment. They aimed at "a clear and thorough description of the various published theories and experiments, regardless of their age". Nor were the invitations evenly distributed among different countries. The list of participants at the Colloquium shows a rather skewed international attendance. The USA topped the list with 22 participants, followed by France with 19, the UK with 7, the USSR with 4, Germany with 3, and Canada, Italy, Japan and the Netherlands with 1 participant, respectively. The primary focus was on the sub-fields of fully developed turbulence and their appropriate representation by the leading experts. Under the authority of a Scientific Executive Committee (Batchelor, Favre, Kovasznay) seven sessions concerning the major branches of turbulence research were established, each one represented by a session president who chose the respective themes and participants (Favre, 1962, pp. 5–13):

Diffusion and Lagrangian Effects (S. Corrsin)
Energy Transfer in Homogeneous Turbulence (G. Batchelor)
Steady Fully Developed Turbulence (R. W. Stewart)
Free Turbulence (H. W. Liepmann)
Turbulent Boundary Layers (H. W. Liepmann, H. Schlichting)
Turbulence in Compressible and Electrically Conducting Media (L. S. Kovasznay)
New Concepts and Recent Contributions (J. Kampé de Fériet, L. S. Kovasznay)

The theme of each session was elaborated in a few lectures by invited speakers, extensively discussed, and finally commented by the session president. The list of invited speakers included not only established authorities like Kolmogorov but also scientists in the early stages of their career; by and large, it reads like a who-is-who in turbulence (Favre, 1962, pp. 469–470).

13.2 Turbulent Diffusion

Diffusion by turbulence (rather than molecular motion) was an old and recurring topic on the agenda of turbulence research. Forty years after Taylor's paper on "Diffusion by Continuous Movements" (see Sect. 7.1) turbulent diffusion has nothing lost of its challenge for the statistical theory of turbulence. Only three years ago Batchelor had chaired a session on the same theme at the IUTAM-IUGG-symposium on "Atmospheric Diffusion and Air Pollution" (see Sect. 12.3). He had extended Taylor's pioneering work with a two-part work on the turbulent diffusion about a fixed and a moving center (Batchelor, 1949, 1952) and surveyed the field together with Townsend (Batchelor and Townsend, 1956). Thus he or Townsend could have authoritatively chaired the first session at the Marseille Colloquium. But Townsend did not travel to Marseille, and Batchelor preferred to reserve for himself the chair for the subsequent session on homogeneous turbulence. Thus the stage was left to Corrsin as chairman for the session on turbulent diffusion.

Corrsin's commitment was not accidental. With his appointment as Assistant Professor of Aeronautics at Johns Hopkins University Kármán's GALCIT in Pasadena had obtained an offshoot for turbulence research in the East of the USA. "Dr. Corrsin is continuing an experimental research program he began at the California Institute of Technology on the nature of turbulent flows and diffusion at velocities small compared with the velocity of sound," Clauser had reported to the dean of his faculty in November 1948. "This research is being carried out under the auspices of the National Advisory Committee for Aeronautics."[7] Corrsin's appointment at Johns Hopkins University was not the least due to his expertise with applications of hot-wire techniques for direct measurements of turbulent diffusion such as in the spreading of heat or in the mixture of gases (Corrsin, 1949). Another fresh addition to the Department of Aeronautics of Johns Hopkins University, Kovasznay, had emigrated from Hungary. Kovasznay improved the hot-wire anenometry for turbulence measurements. Hand in hand with these experiments, "Dr. Francis H. Clauser, department chairman, and Dr. Stanley Corrsin have devised formulae by which they can predict the behavior of turbulence, and more accurately analyse turbulent heat transfer," read a proposal for a research grant to extend these investigations.[8] Together with his first doctoral students—Mahinder Uberoi (1924–2009), Alan L. Kistler (1928–2019), John L. Lumley (1930–2015)—Corrsin established within few years turbulence as a major research focus of Johns Hopkins University (Lumley and Davis, 2003)—with turbulent diffusion as Corrsin's favourite research subject. It was also the subject of close collaboration with Uberoi as his first doctoral student and the gateway into further inquiries about the relation between the Lagrangian and Eulerian description of turbulent diffusion (Uberoi and Corrsin, 1953, pp. 1–2):

> Up to the present time little theoretical or experimental work has been done to find relations, if any, between the Lagrangian statistical measures of a turbulent field with its Eulerian statistical measures. Since turbulence dynamics seems best handled in the latter terms and turbulent diffusion in the former, it is evident that such a connection is important. Hence, one of the purposes of the present experiments has been to compare the magnitudes of some of these quantities under variations in the turbulent field.

In the Eulerian view one considered the turbulent field at a fixed observation point through which the fluid was passing, whereas in the Lagrangian view one was following a moving material point along its path in the fluid. Taylor's pioneering study in 1921 had adopted the latter view, while his statistical theory of turbulence in the 1930s and most other studies that applied to the turbulence in wind tunnels used the Eulerian approach.

The Lagrangian view suggested itself also for the study of free-stream boundaries of turbulent flows such as jets and wakes. On one side of such boundaries the flow

---

[7] Clauser to Kouwenhoven, 10 November 1948. Records of the Department of Aeronautics, The Johns Hopkins University (henceforth abbreviated as Clauser Papers), Box 1a, folder: Biennial Report of the School of Engineering, 1948.

[8] Proposal to Guggenheim Foundation, March 1954. Clauser Papers, Box 1a, folder: Guggenheim Approach.

was turbulent, on the other side of the boundary it was assumed as a turbulence-free potential flow. Corrsin investigated this problem together with Kistler. They named the boundary a "laminar superlayer" and conceived a scheme that suggested how the boundary forms an increasingly wrinkled front due to turbulent diffusion. "It is shown that the rate of increase of wrinkle amplitude of the front can be roughly explained as a Lagrangian diffusion process, using the statistical properties of the turbulence in the fully turbulent zone." (Corrsin and Kistler, 1953, p. 1033). A few years later Corrsin made the unsolved relations between Lagrangian and Eulerian statistics the subject of Lumley's doctoral work (Lumley, 1957). Only in the case of a simple model (a one-dimensional random walk) the Lagrangian and Eulerian correlation functions could be computed and both views reconciled with another (Lumley and Corrsin, 1959). In July 1961 Lumley submitted "An Approach to the Eulerian–Lagrangian Problem" to the *Journal of Mathematical Physics* which displayed common features with other theories such as Feynman's path integral formulation of quantum theory ("sum over histories"). The Eulerian–Lagrangian problem revealed fundamental mathematical problems with random functions, and turbulent diffusion served as a showcase (Lumley, 1962a).

Hence Corrsin engaged Lumley as the first speaker of the session on turbulent diffusion at the Marseille Colloquium with a lecture on "The Mathematical Nature of the Problem of Relating Lagrangian and Eulerian Statistical Functions in Turbulence". The summary provides a glimpse into the nature of this problem (Lumley, 1962b, p. 18):

> The implications of, and relationships between, assumptions of homogeneity, isotropy, stationarity and incompressibility are examined for corresponding Lagrangian and Eulerian fields. It is found that in an incompressible fluid, isotropy and homogeneity of the Eulerian field imply the corresponding property of the Lagrangian field [...] The problem of obtaining the statistics of point trajectories is examined [...] of obtaining the velocity field: on an analytically prescribed path (Eulerian); on a random, but statistically independent path; on a statistically dependent path (Lagrangian). A technique is described [...] for obtaining the functional probability of a point trajectory during a time interval. [...] Attention is given to the problem of integrating the functional probability (in function space) so as to give ordinary densities.

The challenge to relate the two views mathematically to another appeared daunting. But the Eulerian–Lagrangian problem was not merely an issue for theorists. In the second lecture of the session on turbulent diffusion Corrsin reviewed this research field more generally (Corrsin, 1962b). The discussion of this lecture revolved around the question what experiments preferably should be undertaken to provide the required data for further analysis. "Considerably more sophisticated Lagrangian data might be very stimulating," Corrsin pointed to the experimental aspects of the Eulerian–Lagrangian problem at another occasion. But such data were "not too likely unless shortcuts can be devised to avoid the tediousness of extracting data graphically from thousands of particle trajectory photographs" (Corrsin, 1961).

Corrsin's survey of turbulent transport theories ignored molecular diffusion. The third speaker of the session, Philip Geoffrey Saffman (1931–2008), dealt with this lacuna with a lecture on "Some Aspects of the Effects of the Molecular Diffusivity in Turbulent Diffusion". Saffman was a doctoral student of Batchelor and had published in the preceding year a paper on the same subject (Saffman, 1960). He showed that the combined effect of thermal and turbulent motion slowed down the diffusion—contrary to the general view that molecular diffusion rather enhanced turbulent diffusion. In the case of a line source in a field of homogeneous turbulence "the mean-square width of the wake is *less* [Saffman's emphasis] than it would be if the processes of molecular and turbulent diffusion were independent and additive" (Saffman, 1962, p. 54).

The last speaker of this session, Hinze, focused on "Dispersion in Turbulent Shear Flow" (Hinze, 1962a). In his textbook on turbulence, Hinze had dedicated one chapter to "Transport Processes in Turbulent Flows" (Hinze, 1959, Chap. 5). In his Marseille presentation he juxtaposed the Lagrangian view of dispersion in turbulent flows with a constant mean velocity in one direction (as treated for the first time in Taylor, 1922) with other flow configurations (such as in experiments by Batchelor and his collaborators with a suspension of small spheres in a turbulent pipe flow (Batchelor et al., 1955)) or the dispersion from a fixed source in a fully turbulent shear flow (Hinze, 1962b).

Both the discussions of each lecture and the summarizing commentary at the end of the entire session showed that the theory of turbulent diffusion was far from a satisfactory solution. Some hopes were raised into the application of digital computers that had been developed since the end of World War II but were not yet available on a large scale. Corrsin concluded his commentary with the following outlook on future investigations:

> Diffusion is of great intrinsic interest. Some research may be of interest in this area. For example:
>
> (1) More experiments are needed in water and wind tunnels, with a wide range of Reynolds and Prandtl numbers.
>
> (2) Taylor suggested to calculate the dynamic effect of the contaminants used in the experiments.
>
> (3) With the large electronic digital calculators it may be possible to pursue the problem theoretically. For example, one could try the Monte-Carlo method to integrate functionals of probability densities.
>
> (4) Calculators may also be useful for numerical experiments, for example for "random walks" which have both Eulerian and Lagrangian characteristics.
>
> (5) For mathematicians, I suggest to study functions that depend on themselves.[9]

[9] "La diffusion a beaucoup d'intérêt intrinsèque. Certaines recherches pourront etre intéressantes dans ce domaine. Par example:

(1) Il faut faire plus d'expériences dans les souffleries à air et à eau, avec une grande étendue de nombres de reynolds et de Prandtl.

(2) Taylor a suggéré de calculer l'effet dynamique des contaminants utilisés dans les expériences.

(3) Avec les grandes calculatrices digitales électroniques il sera peu-être possible de poursuivre le problème par la voie théorique. Par example, on pourra essayer la méthode "Monte-Carlo" pour intégrer des densités de probabilité fonctionelle.

(4) Les calculatrices pourront aussi être utiles pour des expériences numériques, par exemple pour les "marches aléatoires" qui ont des Charactéristiques à la fois Eulèriennes et Lagrangiennes.

13.3 Homogeneous Turbulence

The theme of the subsequent session was "Energy Transfer in Homogeneous Turbulence"—hand-tailored for Batchelor to chair it and to survey what has been achieved in this field since his revelation of the "remarkable series of coincidences" in K41 (see Sects. 7.5 and 11.1). The objective was "a critical assessment of the progress that has been made towards an understanding of the dynamics of homogeneous turbulence". Homogeneity meant invariance with regard to translations of the frame of reference, "this kind of turbulence being both convenient for experimental study and apparently simpler theoretically in view of the absence of a mean velocity field". The theories referred mostly to grid-generated turbulence in wind tunnels measured by hot-wire instruments. But even the idealisation of homogeneity was insufficient to arrive at a solution of "the dynamical problem" for which Batchelor presented an integral equation for the Fourier transform of the velocity. The non-linear term in this equation prevented a general solution. "After about 20 years of patient inquiry, it is relevant to ask: is it likely that the dynamical problem will ever be solved without the aid of hypotheses?" But he acknowledged that the focus on homogeneous turbulence led to "the firm establishment of statistical terminology" and the current view "that the use of correlation and spectrum functions, etc., is inevitable and that random function theory provides the appropriate analytical framework for a description of turbulent motion" (Batchelor, 1962b, pp. 85–88).

From a more physical vantage point the problem with homogeneous turbulence could be expressed as the question about the distribution of kinetic energy among the eddies of different size. However, the approach that seemed most straight-forward—the expansion in powers of the wave-number—turned out to be "horribly complex", Batchelor remarked with reference to a recent paper co-authored with his doctoral student Ian Proudman (Batchelor and Proudman, 1956); it only amounted to the law about the ultimate decay of the total energy in time according to a certain power law ($\sim t^{-5/2}$). "This is probably a closed chapter in the history of work on homogeneous turbulence" (Batchelor, 1962b, p. 90). In principle the variation of the spectrum of homogeneous turbulence with time could be computed by series expansions to higher degrees of accuracy, but Batchelor doubted whether this approach contributed to a physical understanding of turbulence. Thus he suggested other potential analytical procedures, such as a direct numerical computation of vorticity:

> In these days of high-speed computers there is scope for much more extensive calculations. It would be worth-while, I think, to carry out a numerical integration of the vorticity equation in one realization of a field of homogeneous turbulence, and to verify directly some of the speculations which have been made about the small-scale structure of the turbulence. The numerical task is enormous, but time and the continual improvement of computing machines are on our side here. (Batchelor, 1962b, p. 90)

The small-scale properties were key to Kolmogorov's hypothesis, which assumed that the motion at this scale is in a universal state of equilibrium that is determined

---

(5) Pour les mathématiciens, je suggère d'étudier les fonctions qui dépendent d'elles-mêmes." (Corrsin, 1962a, p. 82).

only by the viscosity ν of the fluid and the rate of energy dissipation per unit mass ε. Twenty years after the birth of K41 this was still "as much of a hypothesis as it was at the beginning". Batchelor expressed his belief in the "intrinsic naturalness" of this theory, but he admitted that he could conceive it being wrong. If so, "the reasons for this would be almost as interesting as the theory itself" (Batchelor, 1962b, p. 93). One still unresolved issue concerned the so-called "flatness factors" of the probability distribution of velocity derivatives. According to K41 the flatness factors should be constant in the limit of high Reynolds numbers, but they showed "a remarkable increase," Batchelor commented with reference to a paper that he had co-authored in 1949 with Townsend. If this feature was confirmed by experiments it suggested "that there is an increasing degree of spottiness in the spatial distribution of energy of small-scale components." In his monograph Batchelor had named this feature "intermittency" (Batchelor, 1953, p. 184). But he called for further measurements "before we can be said to have an understanding of the small-scale structure of turbulence" (Batchelor, 1962b, p. 94). For the time being, doubts about K41 were postponed. A critical test of the theory would have to focus on the rate of transfer of energy across the wave-numbers as outlined in Weizsäcker's and Heisenberg's theories (von Weizsäcker, 1948; Heisenberg, 1948). "It is our business at this colloquium to make a dispassionate assessment of the present state of knowledge of turbulence," Batchelor concluded with a rather sober evaluation of recent progress: "Speaking very broadly, the basic dynamical problem has been in the doldrums, to some extent, during the last 10 years, and has been in need of good ideas and new lines of approach." (Batchelor, 1962b, p. 96).

One such new line of approach was presented by Robert Kraichnan (1928–2008), the second speaker of this session. Kraichnan had started his career in 1949 as a theoretical physicist with a doctoral dissertation on particle physics. He became interested in turbulence when he assisted Albert Einstein (1879–1955) in his quest for a unified field theory. He considered the Navier–Stokes equations as a non-linear field problem where theoretical speculations could be checked by experiment. "After initial surprise that turbulence did not succumb rapidly to field-theoretic attack, I have been trapped ever since," he recalled later about his encounter with the theory of turbulence (quoted in Eyink and Frisch, 2011, p. 330). From the very beginning Kraichnan presented his theory as an alternative to K41:

> The present theory is quite inconsistent with the widely accepted hypothesis of Kolmogorov which holds that at high Reynolds numbers modes of sufficiently high k do not undergo significant direct dynamical interaction with the strongly excited low-k part of the motion, but are in a statistical state determined solely by the rate of energy-cascade [...] As a counter-consideration to Kolmogorov's original argument in the x-space representation, it may be noted that the fine-scale structure of high Reynolds number turbulence consists typically not of compact blobs but of a complicated tangle of *extended* vortex filaments and sheets. (Kraichnan, 1958b, p. 1421)

In a subsequent paper Kraichnan suggested the label "direct-interaction approximation" for his approach (Kraichnan, 1958a, p. 358). The comprehensive account appeared one year later in the *Journal of Fluid Mechanics*, where Kraichnan also duly acknowledged "valuable comments and criticism" of Batchelor, Proudman and others (Kraichnan, 1959). When he presented his theory again in 1961 at the Marseille Colloquium, he set it in perspective to other approaches from the 1950s (Kraichnan, 1962). It became a memorable event and left a strong impression, especially among Batchelor's disciples. Keith Moffatt (born 1935) recalled still fifty years later:

> Kraichnan simply stood at the microphone and spoke for 40 minutes with extraordinary lucidity, and without the help of a single visual aid, even chalk on blackboard! His lecture was followed by a critique of "Kraichnan's theory of turbulence" by Ian Proudman, a younger colleague of Batchelor [...] Proudman's critique was savage by any standard, his conclusion being: "I do not hold out much hope for useful results from the theory [of Kraichnan]". Kraichnan was understandably perturbed by this unfavourable assessment. (Moffatt, 2012, p. 5)

Kraichnan's approach amounted to a wave-number dependence of the energy spectrum proportional to $k^{-3/2}$, "which differs only slightly from the generally accepted Kolmogorov $k^{-5/3}$ law" (Kraichnan, 1962, p. 105). The acceptance of the latter had been largely based on theoretical considerations—until the last speaker of the session, Thomas Herbert Ellison, called attention to recent measurements of turbulent flow in a tidal channel at extremely high Reynolds numbers (up to $3 \cdot 10^8$). "These observations," Ellison argued with reference to slides presented at the Colloquium, "provide what is probably by far the most convincing demonstration of the correctness of the Kolmogoroff theory that has yet been made" (Ellison, 1962, p. 115). The measurements were presented also at the subsequent IUGG-IUTAM-Symposium (Stewart and Grant, 1962) and published in the *Journal of Fluid Mechanics* (Grant et al., 1962). Ellison, a former student of Taylor, pointed also to other evidence which was found "quite commonly in the atmosphere". He also listed theoretical elaborations of Kovasznay, Obukhov and Heisenberg. They "differ little from each other," he summarized their specific results. "Comparison with the measurements by Grant et al. shows fair agreement" (Ellison, 1962, p. 113). Obviously there was strong support of the $k^{-5/3}$ law.

Despite an overall acceptance of K41, Batchelor's summary of the discussion at the end of this session exposed some doubts and demands for future research. Like in the first session, some thought was given to the role of high-speed computers. But Corrsin argued that direct numerical integration of the Navier–Stokes equation with a random initial distribution of velocities was far beyond the reach of existing computers. Contrary to this pessimistic view which referred to the computation of small-scale turbulence at high Reynolds numbers, "Taylor and Liepmann were of the opinion that some useful results might be computable for turbulence of moderate Reynolds number" (Batchelor, 1962a, p. 123). More attention, however, was given to future experimental investigations:

> At several points in the discussion, the need for more measurements of the kinetic energy spectrum and other quantities in turbulence at very high Reynolds number was emphasized. The available theories concerning the small-scale components of homogeneous turbulence are apparently applicable only at Reynolds numbers well above those at which most of the measurements have been made. Kovasznay suggested that the time has come for the investment of a large amount of money in some experimental facility which would make possible measurements of the small-scale components at really high Reynolds numbers. (Batchelor, 1962a, p. 123)

13.4 Enter Kolmogorov

Kolmogorov was absent at the session on homogeneous turbulence. He entered the stage only at the last session of the Colloquium which was dedicated to "New Concepts and Recent Contributions"—with a paper on "Précisions sur la Structure Locale de la Turbulence dans un Fluide Visqueux aux Nombres de Reynolds Élevés". It was printed in the Proceedings in French and in Russian (Kolmogorov, 1962b); an English translation appeared separately in the *Journal of Fluid Mechanics* (Kolmogorov, 1962a). In distinction to his former theory it became labelled as K62. Closely related to Kolmogorov's contribution was a paper by Obukhov on "Some Specific Features of Atmospheric Turbulence" presented in the following week at the IUTAM-IUGG-Symposium (Obukhov, 1962a). When its essential part was republished in the *Journal of Fluid Mechanics* it was put ahead of Kolmogorov's article (Obukhov, 1962b). Obukhov's part in the development of Kolmogorov's ideas, both twenty years ago and again by the time of the Marseille events, has been acknowledged by designating these theories also as KO41 and KO62, respectively (Falkovich, 2011).

Kolmogorov introduced his presentation at the Marseille Colloquium with a brief recollection of Landau's objection that his old theory suffered from an inconsistency: The transfer of energy, ε, could not happen so smoothly from the coarser to the finer vortices because of its random and chaotic character ("le méchanisme de la transmission de l'énergie des tourbillons plus gros vers les plus fine a un charactère fortuit et chaotique"); an increasing ratio between the largest and the smallest eddies would result in an unlimited growth of the spatial fluctuations of ε ("avec l'augmentation du rapport L/l, la variation de la dissipation de l'energie ε [...] doit augmenter d'une facon illimitée"). In other words, Landau had undermined Kolmogorov's old theory with its assumption of a constant ε. The way out of this blow came with a "most recent" approach by Obukhov ("C'est tout dernièrement que A. M. Obukhov avait tracé le chemin ...") which replaced the constant ε by a spatially and temporally varying $\varepsilon_r(x, t)$, where r is a mean radius between the smallest and largest eddies. Based on this new concept Kolmogorov conceived a theory that not only accounted for Landau's objection but also for the concern about intermittency (Kolmogorov, 1962b).

"This lecture, Kolmogorov's final contribution to turbulence, was, to put it mildly, a bit of a bombshell!" Moffatt recalled many years later:

> Kolmogorov delivered his lecture in French, and his only visual aid was the blackboard. What was so startling was that (following reference to parallel work of Obukhov and to comments that he attributed to Landau) Kolmogorov proceeded to point out fundamental inconsistencies in his own theory, in that this theory failed to take account of the spatial fluctuations in the rate of turbulent dissipation of energy ε, the key parameter of the theory. Kolmogorov offered arguments supporting a lognormal probability distribution for this quantity, and showed that the exponent in the energy spectrum would be slightly modified through this "intermittency" effect. [...]

> Kolmogorov's paper marked the onset of a widespread effort in the turbulence research community to understand the phenomenon of intermittency, for which many competing models have been proposed. (Moffatt, 2012, p. 3)

However, the perception of Kolmogorov's lecture as a "bombshell" was rather with hindsight. There is no record of a debate on intermittency in the Proceedings. The chairman of this session, Kampé de Fériet, referred to Kolmogorov's lecture as if it were merely an addition to an established theory:

> Professor Andrei Kolmogorov was kind enough to tell us about the progress of his new research on the theory of local isotropy, which he first formulated in 1941. In particular, he proved that the energy dissipation ε per unit volume tends, when the local scale becomes very small, towards a random variable such that log ε follows the normal law.[10]

Neither Kolmogorov nor Batchelor further explored the implications of the new theory. Although Batchelor and Townsend (who was not attending Marseille '61) had already in 1949 called attention to "intermittent" behaviour ("As the wave-number is increased the fluctuations seem to tend to an approximate on-off, or intermittent, variation" (Batchelor and Townsend, 1949, p. 252)), Batchelor did not play again the role of a missionary for Kolmogorov's theory as he had done fifteen years ago. After Marseille he turned to other research fields. "Given that Batchelor was already frustrated by the mathematical intractability of turbulence, it was perhaps the explicit revelation that all was not well with Kolmogorov's theory that finally led him to abandon turbulence in favour of other fields" (Moffatt, 2011, p. 279).

13.5 Jets, Wakes, Wall-Bounded and Other Turbulent Flows

Intermittency was not merely a theoretical feature of the new Obukhov–Kolmogorov-concept. "The single most important experimental finding in turbulent shear flows is the discovery of the existence of coherent regions of nearly homogeneous vorticity fluctuations separated from the rest of the fluid by sharply defined interfaces", Liepmann introduced the session on "Free Turbulence" at the Marseille Colloquium. He referred to Townsend who had coined in 1949 the term "intermittency" for this phenomenon. "The intermittent occurrence of turbulent regions near the edge of a shear flow [...] demonstrates a large scale superstructure imposed upon a fine scale turbulent flow. It is clear that intermittency is a feature which has to be incorporated into any theory of free turbulence" (Liepmann, 1962b, p. 215). The types of free turbulent flows where this feature was observed were jets, wakes and the mixing of plane flows streaming at different velocities along a common interface. Donald Coles

[10] "le Professeur Andrei Kolmogorov a bien voulu nous exposer le progrès de ses nouvelles recherches sur la théorie de l'isotropie locale, qu'il a formulée en 1941 pour la première fois. Il a prouvé en particulier que la dissipation d'énergie par unité de volume ε tend, lorsque l'échelle locale devient très petite, vers une variable aléatoire telle que log ε suive la loi normale." (de Fériet, 1962, p. 468).

(1924–2013), a collaborator of Liepmann's research group at GALCIT, presented a detailed view on this phenomenon in his lecture on "Interfaces and Intermittency in Turbulent Shear flow" (Coles, 1962). At the end of this session Liepmann expressed the opinion that this feature "removes any hope for a detailed description of free turbulent exchange by a single 'Ansatz' for the transport parameters, but it permits a clearer definition of a number of separate problems for study", such as the structure of turbulent interfaces (Liepmann, 1962a).

Thus the Marseille Colloquium became the stage for presentations from the frontiers of research beyond the statistical theory of homogeneous turbulence. In the session on "Steady Fully Developed Turbulence" Hinze surveyed turbulent pipe flow "in the light of applied measuring techniques of measured distributions of mean values, in particular of the mean-velocity" (Hinze, 1962b). By and large the measurements agreed with the logarithmic "wall law" and Kármán's concept of similarity of fully developed turbulence (see Sect. 6.3). But the evaluation resulted in a considerable variation of the involved constants:

> A closer study of the experimental data now reveals that the scatter of data appears not to be entirely random, but that more or less systematic deviations seem to exist, suggesting that the above concepts might be not strictly correct. These concepts result only in an approximate description of the actual phenomena. So the law of the wall and the velocity-defect law might be only of approximate value, and the von Kármán's constant might be not a constant and Reynolds number similarity might not exist or at the most at much higher Reynolds numbers than assumed hitherto. (Hinze, 1962b, pp. 130–131)

More extensive surveys of wall-bounded turbulence were reserved to the session on "Turbulent Boundary Layers", co-chaired by Liepmann and Schlichting. Schlichting was prevented from attending, but the Göttingen tradition was nevertheless represented at the Marseille Colloquium by the first speaker of this session, Julius C. Rotta (1912–2005), who had started his investigations on turbulence in the footsteps of Prandtl at the Göttingen Max-Planck-Institut für Strömungsforschung (the postwar successor of Prandtl's Kaiser-Wilhelm-Institut für Strömungsforschung). Rotta's expert knowledge on turbulent boundary layer theory was evident, for example, from a report of his institute that was translated in English and widely distributed as a NACA Technical Memorandum (Rotta, 1950, 1953) and from a two-part review on the statistical theory of non-homogeneous turbulence (Rotta, 1951a, b). He also did not shy away from experiments (Rotta, 1956). At the Marseille Colloquium Rotta surveyed a host of recent boundary layer measurements and attempted to provide a coherent view that combined theoretical approaches with experimental data. He concluded his survey with four statements about the current state of boundary layer turbulence:

> (1) The behaviour of the flow in the vicinity of the wall is very different from the flow in the outer part of the boundary layer.
>
> (2) The flow near the wall is determined by local parameters while, in the outer part, the flow is influenced by the upstream history in a very complex manner.
>
> (3) The law of the wall flow has been disclosed by experimental investigations and similarity considerations. [...]

(4) The velocity profile of the outer part can be described to a good approximation in many cases. But many efforts are still required until the outer velocity profile can adequately be related to the pressure distribution. (Rotta, 1962, p. 282)

The next speaker at the boundary layer session was Kistler, who had meanwhile moved to the Jet Propulsion Laboratory, with a lecture on "The Structure of the Laminar Sublayer". The laminar sublayer was regarded as the thin region where the shear stresses are transmitted to the boundary. Theories about this region were mostly "modifications of the mixing length theory", Kistler summarized the progress since the early approaches of Prandtl and Kármán. "These theories have been useful, particularly for heat transfer calculations, but don't shed light on the basic physical processes occurring in the sublayer." But after the review of the most recent theories he arrived at the conclusion that despite qualitative agreement with measurements this was only "a promising start toward a complete theory of the turbulent flow near a solid boundary" (Kistler, 1962, pp. 288, 296). By and large the same observation was made by Alfred Walz from the Deutsche Versuchsanstalt für Luftfahrt, the last speaker of the boundary layer session in a talk on "Compressible Turbulent Boundary Layers". He concluded "that for removing the present difficulties in understanding the discrepancies between experimental results and theoretical or semi-empirical predictions, more interest must be centered on what is happening within the laminar sublayer" (Walz, 1962, p. 346).

The subsequent session on "Turbulence in Compressible and Electrically Conductive Media" was chaired by Kovasznay from the Aeronautics Department at Johns Hopkins University. Kovasznay discerned two distinct modes of turbulence research: an approach that concentrated all efforts on an "oversimplified problem" like K41, and another that takes all kinds of complications into account, even if only "in the large", such as compressibility or electrical conductivity. He regarded the latter approach as "somewhat analogous" with the life sciences where one also did not wait until everything was understood on the most elementary level in atomic physics. Similarly the subjects of Kovasznay's session were investigated before a satisfactory theory of homogeneous turbulence was available. Investigations of turbulence in electrically conducting fluids, i.e. plasmas, were largely spurred by fusion research and astrophysics. The former studied the role of turbulence for the magnetic confinement of hot plasmas, the latter the turbulent flow of cosmical gases (stellar atmospheres, interstellar gas clouds). Both research fields boomed in the 1950s and beyond. Turbulence in compressible media played a role whenever flow velocities came close to or exceeded the speed of sound, such as in supersonic flight. Kovasznay pointed to recent studies concerning the sound at supersonic speeds, where it has been found that the sound energy which is radiated away from turbulent regions was much less than expected. "In low supersonic Mach numbers ($M < 2$) the sound generation is quite small although the sound level on the rather arbitrary scale of the human ear (decibels) may be still quite offensive" (Kovasznay, 1962, p. 361).

The speakers and themes of this session further illustrate what Kovasznay had in mind with the distinction of the two modes of approaches. They all represented the second mode and tackled new kinds of complications. Mark V. Morkovin (1917–

2014) from the Martin Company lectured on the "Effects of Compressibility on Turbulent Flows", John Laufer (1921–1983) from the Jet Propulsion Laboratory on "Sound Radiation from a Turbulent Boundary Layer", and Moffatt on "Turbulence in Conducting Fluids." Moffatt was closely familiar with Batchelor's quest into the most fundamental problems of turbulence. "Now that the extreme complexity of the theory of turbulence in ordinary fluids has been revealed," he echoed Kovasznay, "it may seem to many a rash extravagance to admit to consideration fluids which conduct electricity. The situation is bad enough already—why make it worse by allowing electrons as well as molecules to move unfettered?" (Moffatt, 1962b, p. 396). As an answer Moffatt pointed to the still insufficiently clarified physics of magneto-hydrodynamic turbulence and the related questions raised in astrophysics and plasma physics. But there was also a "perhaps more academic" reason:

> The action of turbulence on a scalar quantity, such as temperature, which is both convected and diffused in the fluid is now fairly well understood. To complete the picture it would be satisfying to understand fully the action of turbulence on a vector quantity which is likewise convected and diffused. The vorticity field is an example, but it is too special, being closely related to the velocity field. The magnetic field in a conducting fluid is the perfect working example. The lines of force of a magnetic field in a fluid of infinite conductivity are convected with the fluid. (Moffatt, 1962b, p. 396)

At the Symposium in the following week Moffat elaborated geophysical consequences of magneto-hydrodynamic turbulence (Moffatt, 1962a). Most presentations of this Symposium were focusing on one or another feature of atmospheric turbulence, but this did not exclude contributions of a more fundamental character such as the presentations of Obukhov and Akiva Yaglom (1921–2007) from the Moscow Institute of Atmospheric Physics of the USSR Academy of Sciences (Obukhov, 1962b; Yaglom, 1962).

One aspect of turbulence, however, was missing both from the sessions at the Colloquium and the Symposium at Marseille: the onset of turbulence. What in the 1920 was regarded as *the* turbulence problem was forty years later considered as a problem of hydrodynamic stability, a speciality in its own right. Meanwhile the transition to turbulence was recognized as a process that was not exhausted by determining one or another hydrodynamic instability. The discovery of "turbulent spots" in otherwise laminar boundary layers (Schubauer and Klebanoff, 1955) hinted at the complexity of the onset of turbulence. "The transition region is marked by the passage of turbulent spots which originate at some point upstream and grow in size as they are carried downstream by the ambient flow", this issue was addressed in a review on "Interfaces and Intermittency in Turbulent Shear Flow" (Coles, 1962, p. 239). Thus the onset of turbulence re-appeared in a different guise: it was no longer a matter of determining critical Reynolds numbers from stability analyses, but of solving the riddle of "intermittency" as a phenomenon of free turbulent flows.

By and large, the variety of themes as well as the applied and academic institutions from which the Marseille contributions originated characterised turbulence as a flourishing research field that transcends established disciplinary boundaries. "About eighty scientists took an active part in the Symposium," Frenkiel remarked about the attendance at this event. Most of them will also have participated in the preceding Colloquium. Never before was turbulent flow subject of such broad and deep scrutiny. It was not too bold to venture the prediction that Marseille '61 would have "great significance to the future progress in turbulence research"—to quote once more one of its organizers (Frenkiel, 1962, p. 3008).

Chapter 14
Epilog

It is beyond the scope of this book to follow the odyssey of turbulence through the decades after Marseille '61. A few observations may suffice to discern some trends.

To begin with the development of the research field of turbulence as a whole. Its consolidation is reflected by new textbooks such as Monin and Yaglom's two-volumes on *Statistical Fluid Mechanics: Mechanics of Turbulence*, first published in Russian in 1965 and 1967, and then in English at the initiative of Lumley (Monin and Yaglom, 1971, 1975). Lumley also published together with co-author Henk Tennekes (1936–2021) at MIT *A First Course in Turbulence* (Tennekes and Lumley, 1972). The latter was praised as a bridge between more elementary introductions on fluid dynamics that "make some mention of turbulence" and the more advanced textbooks "such as those of Batchelor (1953), Townsend (1956), Hinze (1959) or more recently Monin and Yaglom (1971)."[1]

Other indicators of consolidation and growth are conferences and research papers. In addition to the conferences and symposia of IUTAM and other international unions where turbulence was a recurring issue, the European Mechanics Society, founded in 1964, established a series of *European Turbulence Conferences (ETC)* as regular events to take place every other year. Like the role model of Marseille '61 these conferences were devoted "to the fundamental aspects of turbulence of interest to engineers, physicists and mathematicians." The first *ETC* was held in 1986 at Lyon and opened by Batchelor in his capacity as chairman of the European Mechanics Committee. The proceedings list "6 general invited lectures and 57 other contributions, with 152 participants from 19 countries" (Comte-Bellot and Mathieu, 1986, p. VII). Currently these conferences attract "over 450 participants".[2]

In terms of numbers of research papers, turbulence went through a phase of exponential growth. Lumley and Yaglom published in 2001 a review of "A century of

---

[1] Book review in *Journal of Fluid Mechanics*, 58:4, 1973, p. 816.

[2] https://euromech.org/conferences/ETC/ETC18.

M. Eckert, *Turbulence—an Odyssey*, History of Physics,
https://doi.org/10.1007/978-3-030-91459-2_14

turbulence" in which they provided an exponential growth formula for the annual output of papers on turbulence up to 1965 with a doubling period of 11.8 years. The formula was based on the citations in Monin and Yaglom's *Statistical Fluid Mechanics: Mechanics of Turbulence*. Although the numbers from this growth formula should not be taken at face value, they suggest the order of magnitude of the quantitative output of turbulence research: about 100 papers per year by 1965. It is safe to assume that the exponential growth continued for several more years. An exponential growth of scientific papers agrees with the overall tendency in scientometrics in this period (de Solla Price, 1963). Therefore it does not necessarily reflect an exceptional increase of interest in turbulence compared to other sciences. It shows by the mere order of magnitude only that the "new science" of turbulence, like other modern sciences, became since the 1960s a mature research field. However, if the same growth rate would have prevailed until the end of the twentieth century, it would have yielded an output of about 2000 papers per year, which certainly overestimates the actual number. The growth curve must have levelled off. "Still, there are probably more papers in a year than one person can comfortably absorb," the authors of "A century of turbulence" commented on the actual state of their science in the beginning of the twenty-first century (Lumley and Yaglom, 2001, pp. 242–243).

By this time a new *Journal of Turbulence* provided more opportunities for publishing papers on turbulence, in addition to the traditional *Journal of Fluid Mechanics* and *Physics of Fluids*. Another well established journal established in 1949 under the name *Applied Scientific Research* changed its name in 1998 into *Flow, Turbulence and Combustion*. "This apparent narrowing of the journal scope was simply an adjustment to the time and the exponential growth of research output, much in line with the – admittedly extrapolated – Lumley–Yaglom equation for the number of 'turbulence papers'," the editor of this journal remarked about the name change (Hanjalic, 2008, p. 415). The aims and scopes of the *Journal of Turbulence* remind us to the

> fundamental and applied character across the many disciplines in which turbulence plays a vital role. Examples include turbulence arising in engineering fluid dynamics (aerodynamics and hydrodynamics, particulate and multi-phase flows, acoustics, hydraulics, combustion, aeroelasticity, transitional flows, turbo-machinery, heat transfer), geophysical fluid dynamics (environmental flows, oceanography, meteorology), in physics (magnetohydrodynamics and fusion, astrophysics, cryogenic and quantum fluids), and mathematics (turbulence from PDE's, model systems).[3]

In the year 2000 the Navier–Stokes equation, the key to turbulence from a theoretical vantage point, became one of seven Millennium Prize problems in mathematics.[4] Fifty years after Marseille '61 the stage was set once more for discussions on "Fundamental problems of turbulence" at a *Turbulence Colloquium Marseille (TCM2011)*. It was organised in order to celebrate the 50th anniversary of the seminal meeting in 1961. The problems which had been discussed fifty years ago "led to the development

[3] https://www.tandfonline.com/action/journalInformation?show=aimsScope&journalCode=tjot20.

[4] http://www.claymath.org/millennium-problems.

of research areas that are still very much alive to this day," the organizers of *TCM2011* remarked about the progress of their science. They regarded turbulence still as "an open problem since no satisfactory theory is yet available, from either mathematical or physical viewpoint" (Farge, 2013, p. 39). When the issue of mathematics and turbulence was surveyed at *TCM2011*, the reviewer admitted that from this side "the fundamental questions in turbulence" were still lacking an answer (Bardos and Titi, 2013, p. 42).

The persistent topicality of the issues discussed at Marseille '61 can be illustrated by the example of the session on homogeneous turbulence—then chaired by Batchelor, and fifty years later by Batchelor's disciple Moffatt. The pillars of Kolmogorov's theory from 1941 were undermined, the concept of intermittency had surfaced but lacked a common understanding, Kraichnan's "Direct Interaction Approximation" and other issues called for closer scrutiny. "The search for a fundamental theory of turbulence was very much alive then," Moffatt concluded his review on homogeneous turbulence, "it is still alive today, though perhaps in a slightly battered state, since so many approaches have been tried and discarded along the way" (Moffatt, 2012, p. 9).

Similar conclusions were drawn with regard to other issues. "It is a rich field," a survey on turbulent boundary layer research concluded at *TCM2011*, "and the progress in understanding this technically and scientifically important flow has been dramatic, but there is still much to learn" (Wallace, 2013, p. 1). In the same vein the reviewer of pipe and channel flow surveyed much progress over the period of fifty years since Marseille '61. "Nonetheless, this simple pipe or channel flow turbulence has been elusive, refusing to reveal its true nature" (Kim, 2012, p. 16). The reviewers of turbulent jets and wakes also attested "considerable progress" for this field of research, but the experimental discovery of coherent structures in such flows became a new challenge. "An understanding of the mechanics is most likely to be obtained from the vorticity. The subject stands at the beginning of a new era" (Brown and Roshko, 2012, p. 27).

An overall solution to the variety of problems in the different research fields of turbulence seemed elusive. The frustration about this kind of Sisyphus-effort was expressed already in 1979 by Liepmann in an article on "The Rise and Fall of Ideas in Turbulence" with a then famous cartoon about a puzzled looking Dr. Quimsey who had spent years of research to solve a problem, "but now has forgotten the question." In the same vein, the turbulence problem appeared in ever new disguises. "Progress in many directions has been made, indeed significant progress," Liepmann phrased this diversity. "However, the 'turbulence problem' as a whole – whatever that means – remains" (Liepmann, 1979, p. 221).

The rise of new concepts like fractals and the theory of nonlinear dynamical systems ("chaos theory"), the availability of powerful computers and more versatile experimental research facilities evoked new hopes for solving the riddles of turbulent flows (Lumley, 1990). Computational fluid dynamics turned turbulence modelling into a flourishing activity. Direct numerical simulation of turbulence raised hopes to overcome the persistent problem of closure. How the computer has affected recent science in general and turbulence in particular has yet to be investigated historically.

However, a brief look suggests that the new means and methods did not end the recurring ups and downs of high expectations and sobering frustration that also brought changes in the perception of the turbulence problem at different epochs (Eckert, 2019b).

From a sociological vantage point the development of turbulence in the late twentieth century has been characterised as a craving for ordering a research field that was riven with controversy. Turbulence, in particular, does not follow the epistemology of Thomas S. Kuhn (1922–1996) with revolutionary paradigm changes after periods of "normal science":

> So "normal" and Newtonian has this research become that, like a great deal of physics, it now has the deceptive appearance of being routinised, and much of it is conducted under the aegis of engineering. It is firmly ensconced in physics and engineering schools worldwide and has been, in many cases, for over half a century. [...] One might expect that such research would be predominantly theory driven with relatively little controversy. That turbulence research does not fit the interpretation of Kuhn's paradigm as global matrix results in part from difficulties in rendering the phenomena tractable and manipulable. (Turnbull, 1995, p. 13)

Thus the research field of turbulence appears from an epistemological perspective as elusive as "the turbulence problem" from the vantage point of mathematicians and physicists. Despite tremendous progress in a host of applied fields—turbulence still is on a voyage with uncertain outcome: a modern Odyssey.

References

Akerblom F (1908) Recherches sur les courants les plus bas de l'atmosphère au-dessus de Paris. Nova Acta Reg Soc Sci Upsal Ser IV, II(N2)

Alkemade F (1995) Biography. In: Nieuwstadt FTM, Steketee JA (eds) Selected papers of J. M. Burgers. Kluwer, Dordrecht, pp xi–cix

Alkemade F (2019) A century of fluid mechanics in the Netherlands. Springer

Anderson JDJ (1997) A history of aerodynamics and its impact on flying machines. Cambridge University Press, Cambridge

APS (1949) Minutes of the 1948 annual meeting at New York, January 26–29, 1949. Phys Rev 75:1279–1336

Baader J (1797) Theorie des Englischen Zylindergebläses. Neue Phil Abh Bayer Akad Wiss 7:121–168

Baader J (1805) Beschreibung und Theorie des englischen Cylinder-Gebläses, nebst einigen Vorschlägen zur Verbesserung dieser Maschine. Joseph Lindauer, München

Bacon DL, Reid EG (1924) The resistance of spheres in wind tunnels and in air. NACA report, 185

Bagge E, Diebner K, Jay K (1957) Von der Uranspaltung bis Calder Hall. Rowohlt, Hamburg

Barbet L-A (1907) Les grandes eaux de Versailles. H. Dunod et E, Pinat, Paris

Bardos CW, Titi ES (2013) Mathematics and turbulence: where do we stand? J Turbul 14(3):42–76

Barkow E (1917) Windänderung mit der Höhe und Turbulenz. Ann Hydrogr Maritim Meteorol 45:1–6

Barkow E (1924) Der Wind. In: Barkow E, Knoch K (eds) Die Ergebnisse der meteorologischen Beobachtungen der Deutschen Antarktischen Expedition 1911–1912. Veröffentlichungen des Preußischen Meteorologischen Instituts, Springer, Berlin, pp 37–51

Batchelor GK (1946) Double velocity correlation function in turbulent motion. Nature 158(4024):883–884

Batchelor G (1947) Kolmogoroff's theory of locally isotropic turbulence. Proc Camb Philos Soc 43:533–559

Batchelor GK (1949) Diffusion in a field of homogeneous turbulence. Aust J Sci Res Ser A Phys Sci 2:437–450

Batchelor GK (1952) Diffusion in a field of homogeneous turbulence. Proc Camb Philos Soc 48:345–362

Batchelor GK (1953) The theory of homogeneous turbulence. Cambridge University Press, Cambridge

M. Eckert, *Turbulence—an Odyssey*, History of Physics,
https://doi.org/10.1007/978-3-030-91459-2

Batchelor GK (1962a) Discussion de la Séction: Transfert d'Energie en Turbulence Homogène. Mécanique de la Turbulence, Marseille 28 août-2 septembre 1961. Colloq Int Cent Natl Rech Sci 108:123–126

Batchelor GK (1962b) The dynamics of homogeneous turbulence: introductory remarks. Mécanique de la Turbulence, Marseille 28 août-2 septembre 1961. Colloq Int Cent Natl Rech Sci 108:85–97

Batchelor GK (1975) An unfinished dialogue with G. I. Taylor. J Fluid Mech 70(4):625–638

Batchelor G (1981) Preoccupations of a journal editor. J Fluid Mech 106(1):1–25

Batchelor GK (1996) The life and legacy of G. I. Taylor. Cambridge University Press, Cambridge

Batchelor GK, Proudman I (1956) The large-scale structure of homogeneous turbulence. Philos Trans R Soc A 248:369–405

Batchelor GK, Townsend AA (1949) The nature of turbulent motion at large wave-numbers. Proc R Soc A 199:238–255

Batchelor GK, Townsend AA (1956) Turbulent diffusion. Surveys in mechanics. A collection of surveys of the present position of research in some branches of mechanics. Cambridge University Press, Cambridge, pp 352–399

Batchelor GK, Binnie AM, Phillips OM (1955) The mean velocity of discrete particles in turbulent flow in a pipe. Proc Phys Soc Sect B 68(12):1095–1104

Bates CC, Fuller JF (1986) America's weather warriors 18141985. Texas A & M University Press

Battimelli G (1988) The early international congresses of applied mechanics. In: Juhasz S (ed) IUTAM. A short history, Springer, Berlin, pp 9–13

Battimelli G (1996) "Senza alcun vincolo ufficiale": Tullio Levi-Civita a i Congressi Internazionali di Meccanica Applicata. Riv Stor Sci 4(1):51–80

Beaufoy M (1833) Nautical and hydraulic experiments, with numerous scientific miscellanies. In: Henry B (ed) vol 1. South Lambeth, London

Belidor BF (1737) Architecture hydraulique, ou l'art de conduire, d'élever et de ménager les eaux pour les differens besoins de la vie. Jombert, Paris

Benzi R (2011) Lewis Fry Richardson. In: Davidson PA, Kaneda Y, Moffatt K, Sreenivasan KR (eds) A voyage through turbulence. Cambridge University Press, pp 187–208

Bistafa SR (2015) Euler's friction of fluids theory and the estimation of fountain jet heights. Eur Phys J Hist 40:375–384

Blasius H (1908) Grenzschichten in Flüssigkeiten bei kleiner Reibung. Z Angew Math Mech (ZAMM) 56:1–37

Blasius H (1911) Das Ähnlichkeitsgesetz bei Reibungsvorgängen. Phys Z 12:1175–1177

Blasius H (1912a) Das Ähnlichkeitsgesetz bei Reibungsvorgängen. Z Ver Deut Ingen 56:639–643

Blasius H (1912b) Luftwiderstand und Reynoldssche Zahl. (Das Ähnlichkeitsgesetz bei Reibungsvorgängen). Z Flugtech Motorluftschiffahrt 3(3):36–37

Blasius H (1913) Das Ähnlichkeitsgesetz bei Reibungsvorgängen in Flüssigkeiten. Forsch Gebiete Ingenieur 131

Blay M (1986) Recherches sur les forces exercées par les fluides en movement à l'académie royale des sciences: 1668–1669. In: Costabel P (ed) Mariotte, Savant et Philosophe (+ 1684). Analyse d'une renommée. J, Vrin, Paris, pp 91–124

Bobeck P (2006) Henry Darcy in his own words. Hydrogeol J 14:998–1004

Bodenschatz E, Eckert M (2011) Prandtl and the Göttingen school. In: Davidson PA, Kaneda Y, Moffatt K, Sreenivasan KR (eds) A voyage through turbulence. Cambridge University Press, Cambridge, pp 40–100

Boltze E (1908) Grenzschichten an Rotationskörpern in Flüssigkeiten mit kleiner Reibung. PhD thesis, Universität Göttingen

Bossut C (1771) Traité élémentaire d'hydrodynamique, vol 2. Jombert, Paris

Bossut C, le Rond d'Alembert J, de Condorcet M (1777) Nouvelles expériences sur la résistance des fluides. Presented to the academy of science on 17 April 1776. Claude-Antoine Jombert, Paris

Boussinesq J (1877) Essai sur la Théorie des Eaux Courantes. Imprimerie Nationale, Paris

Brandstetter T (2006) Kräfte messen: Die Maschine von Marly und die Kultur der Technik 1680–1840. PhD thesis, Bauhaus Universität Weimar, Weimar

Breford and Möller E (1943) Messungen am Originalflügel des Baumusters P-51 'Mustang'. Deut Luftfahrtforschung Forschungsbericht 1724-2. ZWB

Brown GO (2002a) Henry Darcy and the making of a law. Water Resour Res 38(7):1101–1112

Brown GO (2002b) The history of the Darcy-Weisbach equation for pipe flow resistance. In: Fredrich A, Rogers J (eds) Environmental and water resources history. American Society of Civil Engineers, Reston, VA, pp 34–43

Brown GO (2003) Henry Darcy's perfection of the pitot tube. In: Brown GO, Garbrecht JD, Hager WH (eds) Henry P. G. Darcy and other pioneers in hydraulics. Contributions in celebration of the 200th birthday of Henry Philibert Gaspard Darcy. ASCE, Reston, VA, pp 14–23

Brown DK (2006) The way of a ship in the midst of the sea: the life and work of William Froude. Periscope Publishing

Brown GL, Roshko A (2012) Turbulent shear layers and wakes. J Turbul 13(51):1–32

Burgers JM (1925) The motion of a fluid in the boundary layer along a plane smooth surface. In: Burgers JM, Biezeno CB (eds). Proceedings of the first international congress for applied mechanics, Delft, pp 113–128

Burgers JM, van de Hulst HC (eds) (1949) Problems of cosmical aerodynamics. In: Proceedings of the symposium on the motion of gaseous masses of cosmical dimensions held at Paris, August 16–19, Dayton, OH. Central Air Documents Office (Army-Navy-Air Force)

Calero JS (2008) The genesis of fluid mechanics 1640–1780. Springer

Cassidy DC (2009) Beyond uncertainty: Heisenberg, quantum physics, and the bomb. Bellevue Literary Press, New York

Chandrasekhar S (1961) Hydrodynamic and hydromagnetic stability. Clarendon Press, London

Chentsov NN (1990) The unfathomable influence of Kolmogorov. Ann Stat 1(3):987–998

Clarke J, Geikie A (1910) Physical science in the time of Nero; being a translation of the Quaestiones naturales of Seneca. Macmillan and Company Limited, London

Clauser FH (1957) The structure of turbulent shear flow. Nature 179:60

Coen DR (2007) Vienna in the age of uncertainty: science, liberalism, and private life. University of Chicago Press, Chicago

Coles D (1962) Interfaces and intermittency in turbulent shear flow. Mécanique de la turbulence, Marseille 28 août-2 septembre 1961. Colloq Int Cent Natl Rech Sci 108:230–250

Comte-Bellot G, Mathieu J (eds) (1986) Advances in turbulence. In: Proceedings of the first European turbulence conference, Lyon, 1–4 July 1986. Springer, Berlin, Heidelberg

Corrsin S (1943) Investigation of flow in an axially symmetrical heated jet of air. NACA advance confidential report, 3L23

Corrsin S (1949) Extended applications of the hot-wire anemometer. NACA Tech Note 1864:1–62

Corrsin S (1961) Turbulent flow. Am Sci 49(3):300–325

Corrsin S (1962a) Commentaire de la Section: Diffusion et Point de Vue de Lagrange. Mécanique de la Turbulence, Marseille 28 août-2 septembre 1961. Colloq Int Cent Natl Rech Sci 108:81–82

Corrsin S (1962b) Theories of turbulent dispersion. Mécanique de la Turbulence, Marseille 28 août-2 septembre 1961. Colloq Int Cent Natl Rech Sci 108:27–52

Corrsin S, Kistler AL (1953) Free-stream boundaries of turbulent flows. NACA Rep 1244:1033–1064

Coulomb C (1800) Experiences destinées à déterminer la cohesion des fluides et les lois de leur résistance dans les mouvements très lents (1798). Inst Natl Sci Arts Mém Sci Math Phys 3:246–305

Couplet P (1732) Recherches sur le mouvement des eaux. Hist Acad R Sci 113–168

Crewe ME (2009) The MET office grows up: in war and peace. Published by The Royal Meteorological Society's History of Meteorology and Physical Oceanography Special Interest Group. Published online only: https://www.rmets.org/sites/default/files/papers/hist08.pdf

Damljanović D (2012) Gustave Eiffel and the wind: a pioneer in experimental aerodynamics. Sci Tech Rev 62(3–4):3–13

Darcy H (1856) Les Fontaines Publiques de la Ville de Dijon. Dalmont, Paris

Darcy H (1857) Recherches Expérimentales Relatives au Mouvement de l'Eau dans les Tuyaux. Mallet-Bachelier, Paris
Darcy H, Bazin H (1865) Recherches hydrauliques. Première partie: Recherches experimentales sur l'écoulement de l'eau dans les canaux découverts. Deuxième partie: Recherches experimentales relatives aux remous et à la propagation des ondes. Imprimerie Impériale, Paris
Darrigol O (1998) From organ pipes to atmospheric motions: Helmholtz on fluid mechanics. Hist Stud Phys Biol Sci 29(1):1–51
Darrigol O (2005) Worlds of flow. Oxford University Press, Oxford
Darrigol O (2017) Joseph Boussinesq's legacy in fluid mechanics. C R Mecanique 345(7):427–445
d'Aubuisson de Voisins JF (1828) Sur la résistance que l'air éprouve dans des tuyaux de conduite, faites aux mines de Rancié, en 1825. Ann Mines 367–486
d'Aubuisson de Voisins JF (1834) Traité d'Hydraulique a l'Usage des Ingenieurs. Levrault, Paris
Davidson PA, Kaneda Y, Moffatt K, Sreenivasan KR (eds) (2011) A voyage through turbulence. Cambridge University Press, Cambridge
Davis SH, Pedley TJ (2006) Editorial: JFM at 50. J Fluid Mech 554:1–4
de Fériet JK (1962) Commentaire de la Section Nouvelles Conceptions et Contributions Récentes. Mécanique de la Turbulence, Marseille 28 août-2 septembre 1961. Colloq Int Cent Natl Rech Sci 108:467–468
de Fontenelle BLB (1722) Éloge de M. Couplet. Mém Acad R Sci 124–128
de Fouchy G (1774) Éloge de M. Pitot. Hist Acad R Sci 73:143–157
de Prony G (1804) Recherches physico-mathematiques sur la théorie des eaux courantes. Imprimerie Imperiale, Paris
de Solla Price DJ (1963) Little science, big science. Columbia University Press, New York
Descamps P (2003) D'exorbitants besoins en eau. Sciences et Techniques des Bâtisseurs de Versailles. Cah Sci Vie 74:84–89
Doel RE, Hoffmann D, Krementsov N (2005) National states and international science: a comparative history of international science congresses in Hitler's Germany, Stalin's Russia, and Cold War United States. Osiris, 2nd Series. Politics and science in wartime: comparative international perspectives on the Kaiser Wilhelm Institute 20:46–76
Douglas CKM (1916) Weather observations from an aeroplane. J Scott Meteorol Soc 17(33):65–73
Douglas CKM (1919) Meteorology and aviation. Nature 102(2572):473–476
Dove HW (1857) Über das Gesetz der Stürme. Dietrich Reimer, Berlin
Dryden HL (1931) Reduction of turbulence in wind tunnels. NACA Rep 392:557–565
Dryden HL (1936) Air flow in the boundary layer near a plate. NACA Rep 562:339–364
Dryden HL (1937) The theory of isotropic turbulence. J Aeronaut Sci 4(7):273–280
Dryden HL (1939) Turbulence investigations at the National Bureau of Standards. In: Den Hartog JP, Peters H (eds) Proceedings of the fifth international congress on applied mechanics, Cambridge MA. Wiley, New York, pp 362–368
Dryden HL (1947) The international congress for applied mechanics. Science 105(2720):167–169
Dryden HL (1955) Fifty years of boundary-layer theory and experiment. Science 121(3142):375–380
Dryden HL, Heald RH (1926) Investigation of turbulence in wind tunnels by a study of the flow about cylinders. NACA Rep 231
Dryden HL, Kuethe AM (1929a) Effect of turbulence in wind tunnels measurements. NACA Rep 342
Dryden HL, Kuethe AM (1929b) The measurement of fluctuations of air speed by the hot-wire anemometer. NACA Rep 320
Dryden HL, Schubauer GB Jr, Mock WC, Skramstad HK (1936) Measurements of intensity and scale of wind-tunnel turbulence and the relation to critical Reynolds number of spheres. NACA Rep 581:109–140
du Buat P (1786) Principes d'Hydraulique, Vérifiés par un grand nombre d'Expériences faites par ordre du Gouvernement. Monsieur, Paris
Durand WF (1898) The resistance and propulsion of ships. Wiley, New York

Eberhard P, Juhasz S (eds) (2016) IUTAM: a short history, 2nd edn. Springer, Open Access
Eckert M (2006) The dawn of fluid dynamics. Wiley-VCH, Weinheim
Eckert M (2010) The troublesome birth of hydrodynamic stability theory: Sommerfeld and the turbulence problem. Eur Phys J Hist 35(1):29–51
Eckert M (2017) Ludwig Prandtl—Strömungsforscher und Wissenschaftsmanager. Springer, Berlin, Heidelberg. English translation: Ludwig Prandtl—a life for fluid mechanics and aeronautical research. Springer
Eckert M (2019a) Strömungsmechanik zwischen Mathematik und Ingenieurwissenschaft: Felix Kleins Hydrodynamikseminar 1907/08. Hamburg University Press, Hamburg
Eckert M (2019b) The turbulence problem—a persistent riddle in historical perspective. In: Springer briefs in history of science and technology. Springer, Cham
Eiffel G (1907) Recherches expérimentales sur la résistance de l'air exécutées à la tour Eiffel. L. Maretheux, Paris
Eiffel G (1912) Sur la résistance des sphères dans l'air en mouvement. C R 155:1597–1599
Ellison TH (1962) The universal small-scale spectrum of turbulence at high Reynolds number. Mécanique de la Turbulence, Marseille 28 août-2 septembre 1961. Colloq Int Cent Natl Rech Sci 108:113–121
Emrich R, Klebanoff PS, Elsasser WM, Polachek H (1987) Obituaries: Francois N. Frenkiel. Phys Today 123–124
Exner F (1912) Dynamische Meteorologie. Enzykl Math Wiss VI 1B(3):179–234
Eyink G, Frisch U (2011) Robert H. Kraichnan. In: Davidson PA, Kaneda Y, Moffatt K, Sreenivasan KR (eds) A voyage through turbulence. Cambridge University Press, Cambridge, pp 329–372
Eyink GL, Sreenivasan KR (2006) Onsager and the theory of hydrodynamic turbulence. Rev Mod Phys 78(1):87–135
Eytelwein JA (1801) Handbuch der Mechanik fester Körper und der Hydraulik. Lagarde, Berlin
Falkovich G (2011) The Russian school. In: Davidson PA, Kaneda Y, Moffatt K, Sreenivasan KR (eds) A voyage through turbulence. Cambridge University Press, Cambridge, pp 209–237
Farge M (2013) Foreword: turbulence colloquium Marseille 2011. J Turbul 14(9):39–42
Favre A (1962) Introduction. Mécanique de la Turbulence, Marseille 28 août-2 septembre 1961. Colloq Int Cent Natl Rech Sci 108:5–7
Ferreiro LD (2007) Ships and science. The birth of naval architecture in the scientific revolution, 1600–1800. MIT Press, Cambridge, MA
Ferreiro LD (2020) Bridging the seas: the rise of naval architecture in the industrial age. MIT Press, Cambridge, MA
Flachsbart O (1932) Geschichte der experimentellen Hydro- und Aeromechanik, insbesondere der Widerstandsforschung. Handbuch Exp Phys 4(2):1–61
Föppl O (1911) Windkräfte an ebenen und gewölbten Platten. Jahr Motorluftschiff-Studien Gesellsch 4:51–119
Föppl O (1912) Ergebnisse der aerodynamischen Versuchsanstalt von Eiffel, verglichen mit den Göttinger Resultaten. Z Flugtech Motorluftsch 3:118–121
Forchheimer P (1905) Hydraulik. Enzykl Math Wiss IV(20):324–472
Forman P (2007) Die Naturforscherversammlung in Nauheim im September 1920: Eine Einführung in das Wissenschaftsleben der Weimarer Republik. In: Hoffmann D, Walker M (eds) Physiker zwischen Autonomie und Anpassung. Die Deutsche Physikalische Gesellschaft im Dritten Reich, Wiley-VCH, Weinheim, pp 29–58
Freeze RA (1994) Henry Darcy and the fountains of Dijon. Ground Water 32(1):23–30
Frenkiel FN (1962) Introduction to the proceedings of a symposium on fundamental problems in turbulence and their relation to geophysics. Organized by the International Union of Geodesy and Geophysics and the International Union of Theoretical and Applied Mechanics held in Aix-Marseilles, September 4–9, 1961. J Geophys Res 67(8):3007–3010
Frenkiel FN, Sheppard PA (eds) (1959) Atmospheric diffusion and air pollution. Academic Press, New York, London

Friedlander SK, Topper L (eds) (1961) Turbulence: classic papers on statistical theory. Interscience Publishers, New York and London
Friedman RM (1989) Appropriating the weather: Vilhelm Bjerknes and the construction of a modern meteorology. Cornell University Press, Ithaca, NY
Frisch U (1995) Turbulence. The legacy of A. N. Kolmogorov. Cambridge University Press, Cambridge
Frisinger HH (1977) The history of meteorology: to 1800. Science History Publications, New York
Fritsch W (1928) Der Einfluß der Wandrauhigkeit auf die turbulente Geschwindigkeitsverteilung in Rinnen. Z Angew Math Mech (ZAMM) 8:199–216
Fromm K (1923) Strömungswiderstand in rauhen Rohren. Abhandlungen Aerodyn Inst Tech Hochschule Aachen 3:1–20
Froude W (1872) Experiments on the surface-friction experienced by a plane moving through water. Report of the forty-second meeting of the British Association for the Advancement of Science, pp 118–124
Fuhrmann G (1912) Theoretische und experimentelle Untersuchungen an Ballonmodellen. PhD thesis, Universität Göttingen. Published in: Jahrbuch der Motorluftschiff-Studiengesellschaft, 1911–1912, S. 64–123
Galilei G (1950) Dialogues concerning two new sciences. Dover, New York. Based on Galileo's original Italian text published in 1638
Garber E (1976) Thermodynamics and meteorology (1850–1900). Ann Sci 33(1):51–65
Gebers F (1906) Die Versuchsanstalt "Übigau". Verlag Schiffbau, Berlin
Gebers F (1908) Ein Beitrag zur experimentellen Ermittlung des Wasserwiderstandes gegen bewegte Körper. PhD thesis, Technische Hochscule Dresden
Gericke H (1972) 50 Jahre GAMM. Springer, Berlin
Girard PS (1803) Rapport à l'Assemblée des Ponts et Chaussée sur le projet générale du Canal de l'Ourcq. L'Imprimerie de la République, Paris
Girard PS (1821) Sur l'écoulement uniforme de l'air atmosphérique et du gaz hydrogène carboné, dans des tuyaux de conduite. Ann Chim Phys 16:129–152. Lu à l'Académie des Sciences le 12 juillet 1819
Gisonni C (2003) Henry Darcy and the pipe flow formula. In: Brown GO, Garbrecht JD, Hager WH (eds) Henry P. G. Darcy and other pioneers in hydraulics. Contributions in celebration of the 200th birthday of Henry Philibert Gaspard Darcy. ASCE, Reston, VA, pp 24–36
Görtler H (1940a) Über den Einfluß der Wandkrümmung auf die Entstehung der Turbulenz. Z Angew Math Mech (ZAMM) 20:138–147
Görtler H (1940b) Über eine dreidimensionale Instabilität laminarer Grenzschichten an konkaven Wänden. Nachrichten der Gesellschaft der Wissenschaften zu Göttingen. Mathematisch-physikalische Klasse, Neue Folge, pp 1–26
Görtler H (1941) Instabilität laminarer Grenzschichten an konkaven Wänden gegenüber gewissen dreidimensionalen Störungen. Z Angew Math Mech (ZAMM) 21(250–252):384
Görtler H (1948) Turbulenz. Naturforschung und Medizin in Deutschland 1939–1946. Deut bestimmte Ausg FIAT Rev Ger Sci 5(III):75–100
Grant HL, Stewart RW, Moilliet A (1962) Turbulence spectra from a tidal channel. J Fluid Mech 12:241–268
Greenaway F (1996) Science international: a history of the international council of scientific unions. Cambridge University Press, Cambridge
Guldberg CM, Mohn H (1876) Études sur les Mouvements de l' Atmosphère. Brögger, Christiania
Hadley G (1735) Concerning the cause of the general trade-winds. Philos Trans R Soc 39:58–62
Hagen G (1839) Über die Bewegung des Wassers in engen cylindrischen Röhren. Ann Phys 46:423–442
Hagen G (1855) Über den Einfluss der Temperatur auf die Bewegung des Wassers in Röhren. Math Abh Königlichen Akad Wiss Berlin 1854:17–98
Hager WH (1994) Die historische Entwicklung der Fliessformel. Sch Ing Archit 112(9):123–133
Hager WH (2003) Blasius: a life in research and education. Exp Fluids 34:566–571

Hahn H, Herglotz G, Schwarzschild K (1904) Über das Strömen des Wassers in Röhren und Kanälen. Z Math Phys 51:411–426

Halley E (1687) An historical account of the trade winds, and monsoons, observable in the seas between and near the tropicks, with an attempt to assign the physical cause of the said winds. Philos Trans R Soc 16:153–168

Hammerl C, Lenhardt W, Steinacker R, Steinhauser P (2001) Die Zentralanstalt für Meteorologie und Geodynamik 1851–2001: 150 Jahre Meteorologie und Geophysik in Österreich. ZAMG

Hanjalic K (2008) From "applied scientific research" to "flow, turbulence and combustion": 60th anniversary of a lasting legacy. Flow Turbul Combust 80:413–417

Hansen M (1928) Die Geschwindigkeitsverteilung in der Grenzschicht an einer eingetauchten Platte. Z Angew Math Mech (ZAMM) 8:185–199

Hansen JR (1987) Engineer in charge. A history of the Langley Aeronautical Laboratory, 1917–1958. NASA, Washington. Available at http://history.nasa.gov/SP-4305/sp4305.htm

Harper KC (2006) Meteorology's struggle for professional recognition in the USA (1900–1950). Ann Sci 63(2):179–199

Hartog JPD, Peters H (eds) (1939) Proceedings of the fifth international congress for applied mechanics. Wiley, New York. Held at Harvard University and the Massachusetts Institute of Technology, Cambridge, MA, 12–16 Sept 1938

Heisenberg W (1922) Nichtlaminare Lösungen der Differentialgleichungen für reibende Flüssigkeiten. In: von Kármán T, Leci-Civita T (eds) Vorträge aus dem Gebiete der Hydro- und Aerodynamik (Innsbruck 1922). Springer, Berlin, pp 139–142

Heisenberg W (1924) Über Stabilität und Turbulenz von Flüssigkeitsströmen. Ann Phys 74:577–627

Heisenberg W (1948) Zur statistischen Theorie der Turbulenz. Z Phys 124:628–657

Heisenberg W (1950) On the stability of laminar flow. In: Proceedings of the international congress of mathematicians, vol II. American Mathematical Society, Providence, pp 292–296

Heisenberg W (1951) On stability and turbulence of fluid flows. Technical memorandum NACA TM 1291, NACA. Translated from Ann Phys 74(1924):577–627

Helmholtz H, v Piotrowski G (1860) Über Reibung tropfbarer Flüssigkeiten. Sitzungsber Math Naturwiss Cl Kaiserl Akad Wiss Wien XL:607–658

Herschel C (1897) On the origin of the Chézy formula. J Assoc Eng Soc 18:363–369

Heyman J (1976) Couplet's engineering memoirs, 1726–33. Hist Technol 1:21–44

Hiemenz K (1911) Die Grenzschicht an einem in den gleichförmigen Flüssigkeitsstrom eingetauchten geraden Kreiszylinder. PhD thesis, Universität Göttingen

Hinze JO (1959) Turbulence. An introduction to its mechanism and theory, McGraw-Hill, New York

Hinze JO (1962a) Dispersion in turbulent shear flow. Mécanique de la Turbulence, Marseille 28 août–2 septembre 1961. Colloq Int Cent Natl Rech Sci 108

Hinze JO (1962b) Turbulent pipe-flow. Mécanique de la Turbulence, Marseille 28 août-2 septembre 1961. Colloq Int Cent Natl Rech Sci 108:129–165

Holstein H (1946) The large AVA-tunnel of low turbulence. Rep Transl 83. ZWB

Hopf L (1910) Hydrodynamische Untersuchungen: Turbulenz bei einem Flusse. Über Schiffswellen. Inaugural-Dissertation. Barth, Leipzig

Hopf L (1923) Die Messung der hydraulischen Rauhigkeit. Z Angew Math Mech (ZAMM) 3(5):329–339

Hopf L (1927) Zähe Flüssigkeiten. Handb Phys 7:91–172

Hult J, Hoff NJ (2016) Post world war II activities: from Paris to Lyngby. In: Peter E, Stephen J (eds) IUTAM. A short history, 2nd edn. Springer, pp 31–35

Hunt JCR (1998) Lewis Fry Richardson and his contributions to mathematics, meteorology, and models of conflict. Ann Rev Fluid Mech 30(1):xiii–xxxvi

Hutton C (1812a) Resistance of the air determined by the whirling-machine. Tracts Math Philos Subj Charles Hutton Lond 3:163–208

Hutton C (1812b) Theory and practice of gunnery, as dependent on the resistance of the air. Tracts Math Philos Subj Charles Hutton Lond 3:209–315

Iben O (1880) Druckhöhen-Verlust in geschlossenen eisernen Rohrleitungen. In: Denkschrift des Verbandes deutscher Architekten- und Ingenieur-Vereine. Otto Meissner, Hamburg

IUTAM (ed) (1948) Proceedings of the seventh international congress for applied mechanics, 5–11 Sept 1948. Published by the Organizing Committee. Introduction, vol I, II, parts 1 and 2, vol III, IV. IUTAM

Jackson D, Launder B (2007) Osborne Reynolds and the publication of his papers on turbulent flow. Ann Rev Fluid Mech 2007:19–35

Juhasz S (ed) (1988) IUTAM. A short history, Springer, Berlin

Kaldewey D, Schauz D (eds) (2018) Basic and applied research. The language of science policy in the twentieth century, New York, Oxford, Berghahn

Kevles DJ (1978) The physicists: the history of a scientific community in modern America. Alfred E. Knopf

Kim J (2012) Progress in pipe and channel flow turbulence, 1961–2011. J Turbul 45:1–19

Kistler AL (1962) The structure of the laminar sublayer. Mécanique de la turbulence, Marseille 28 août-2 septembre 1961. Colloq Int Cent Natl Rech Sci 108:287–297

Klein F (1910) Über die Bildung von Wirbeln in reibungslosen Flüssigkeiten. Z Math Phys 259–262

Knibbs GH (1897) On the steady flow of water in uniform pipes and channels. J Proc R Soc New South Wales 31:314–355

Kolmogorov AN (1962a) A refinement of previous hypotheses concerning the local structure of turbulence in a viscous incompressible fluid at high Reynolds number. J Fluid Mech 13:82–85

Kolmogorov AN (1962b) Précisions sur la Structure Locale de la Turbulence dans un Fluide Visqueux aux Nombres de Reynolds Élevés. Mécanique de la Turbulence, Marseille 28 août-2 septembre 1961. Colloq Int Cent Natl Rech Sci 108:447–458

Kolmogorov AN (1991a) Dissipation of energy in the locally isotropic turbulence. Proc R Soc Lond A 434:15–17. First published in Russian in Dokl Akad Nauk SSSR 32(1) (1941)

Kolmogorov AN (1991b) The local structure of turbulence in incompressible viscous fluid for very large Reynolds numbers. Proc R Soc Lond A 434:9–13. First published in Russian in Dokl Akad Nauk SSSR 30(4) (1941)

Kopfermann, Breford (1943) Umschlagpunktmessungen am Originalflügel des Baumusters P-51 "Mustang". Deut Luftfahrtfor Untersuchungen Mitteil (UM) 2035. ZWB

Kovasznay LG (1962) Turbulence in compressible and electrically conductive media. Mécanique de la turbulence, Marseille 28 août-2 septembre 1961. Colloq Int Cent Natl Rech Sci 108:357–365

Kraichnan RH (1958a) Higher order interactions in homogeneous turbulence theory. Phys Fluids 1:358–359

Kraichnan RH (1958b) Irreversible statistical mechanics of incompressible hydromagnetic turbulence. Phys Rev 109(5):1407–1422

Kraichnan RK (1959) The structure of isotropic turbulence at very high Reynolds numbers. J Fluid Mech 5(4):497–543

Kraichnan RH (1962) Relationships among some deductive theories of turbulence. Mécanique de la turbulence, Marseille 28 août-2 septembre 1961. Colloq Int Cent Natl Rech Sci 108:99–106

Kraus H (2008) Grundlagen der Grenzschicht-Meteorologie. Springer, Berlin, Heidelberg

Kutzbach G (1979) The thermal theory of cyclones: a history of meteorological thought in the nineteenth century. Historical monograph series. American Meteorological Society, Boston, MA

Landau LD (1944) On the problem of turbulence. In: Ter Haar D (ed) Collected papers of L. D. Landau. Pergamon Press, Oxford, pp 387–391 (1965)

Launder B, Jackson D (2011) Osborne Reynolds: a turbulent life. In: Davidson PA, Kaneda Y, Moffatt K, Sreenivasan KR (eds) A voyage through turbulence. Cambridge University Press, Cambridge, pp 1–39

Leggett D (2013) Replication, replacing and naval science in comparative context, c. 1868–1904. Br J Hist Sci 46(1):1–21

Lehot M, Désormes M, Clément M (1811) Experiences sur la résistance que le mouvement de l'air éprouve dans les tuyaux d'une grande longeur. J Phys Chim Hist Nat Arts LXXIII:36–40

Leonard A, Peters N (2011) Theodore von Kármán. In: Davidson PA, Kaneda Y, Moffatt K, Sreenivasan KR (eds) A voyage through turbulence. Cambridge University Press, Cambridge, pp 101–126

LePage WL, Nichols JT (1924) The effect of wind tunnel turbulence upon the forces measured on models. NACA Tech Notes 191:1–10

Liepmann H (1943) Investigations on laminar boundary layer stability and transition on curved boundaries. NACA advance confidential report, ACR 3H30

Liepmann H (1962a) Commentaire de la Section: Turbulence Libre. Mécanique de la Turbulence, Marseille 28 août–2 septembre 1961. Colloq Int Cent Natl Rech Sci 108:251

Liepmann H (1962b) Free turbulent flows. Mécanique de la Turbulence, Marseille 28 août-2 septembre 1961. Colloq Int Cent Natl Rech Sci 108:211–227

Liepmann HW (1979) The rise and fall of ideas in turbulence. Am Sci 67(2):221–228

Liepmann HW (1997) Boundary layer transition, the early days. Appl Mech Rev 50(2):R1–R4

Lilienthal O (1889) Der Vogelflug als Grundlage der Fliegekunst: ein Beitrag zur Systematik der Flugtechnik. R. Gaertners Verlagsbuchhandlung, Berlin

Lin C-C (1945) On the stability of two-dimensional parallel flows. Q Appl Math 3:117–142

Lin C-C (1946) On the stability of two-dimensional parallel flows. Q Appl Math 4:277–301

Lin C-C (1955) The theory of hydrodynamic stability. Cambridge University Press

Linke F (1917) Die meteorologische Ausbildung des Fliegers. Oldenbourg, München, Berlin

Lorentz HA (1897) Over den weerstand dien een vloeistofstroom in eene cilindrische buis ondervindt. Versl K Akad Wet Amsterdam 6:28–49

Love AEH (1901a) Hydrodynamik. I. Physikalische Grundlegung. Enzykl Math Wiss IV(15):48–83

Love AEH (1901b) Hydrodynamik. II. Theoretische Ausführungen. Enzykl Math Wiss IV(16):84–147

Lumley JL (1957) Some problems connected with the motion of small particles in turbulent fluid. PhD thesis, Johns Hopkins University

Lumley JL (1962a) An approach to the Eulerian-Lagrangian problem. J Math Phys 3(2):309–312

Lumley JL (1962b) The mathematical nature of the problem of relating Lagrangian and Eulerian statistical functions in turbulence. Mécanique de la Turbulence, Marseille 28 août-2 septembre 1961. Colloq Int Cent Natl Rech Sci 108:17–26

Lumley JL (ed) (1990) Whither turbulence? Turbulence at the crossroads. In: Proceedings of a workshop held at Cornell University, Ithaca, NY, 22–24 Mar 1989. Lecture notes in physics, vol 357. Springer

Lumley JL, Corrsin S (1959) Random walk with both Lagrangian and Eulerian statistics. Adv Geophys 6:179–183

Lumley JL, Davis SH (2003) Stanley Corrsin: 1920–1986. Ann Rev Fluid Mech 35:1–10

Lumley JL, Yaglom AM (2001) A century of turbulence. Flow Turbul Combust 66:241–286

Lynch P (2006) The emergence of numerical weather prediction: Richardson's dream. Cambridge University Press, Cambridge

Macagno EO (1986) Leonardian fluid mechanics in the Codex Atlanticus: general survey. Iowa Institute for Hydraulic Research, Iowa City, IA

Maffioli CS (1994) Out of Galileo. The science of waters, 1628–1718. Erasmus Publishing, Rotterdam

Magnusson RJ (2001) Water technology in the middle ages. The Johns Hopkins University Press, Baltimore and London

Mariotte E (1686) Traité du Mouvement des Eaux et des Autres Corps Fluides. Estienne Michallet, Paris

Marusic I, Broomhall S (2021) Leonardo da Vinci and fluid mechanics. Ann Rev Fluid Mech 53(1):1–25

Masters RD (1998) Fortune is a river: Leonardo Da Vinci and Niccolò Machiavelli's magnificent dream to change the course of Florentine history. The Free Press, New York

Meier GEA, Sreenivasan KR (2004) IUTAM symposium on one hundred years of boundary layer research. Springer, Dordrecht

Mekysn D (1946) Asymptotic integrals of a fourth order differential equation containing a large parameter. Proc Lond Math Soc s2-49(1):436–457

Meneveau C, Riley JJ (2011) Stanley Corrsin. In: Davidson PA, Kaneda Y, Moffatt K, Sreenivasan KR (eds) A voyage through turbulence. Cambridge University Press, Cambridge, pp 238–275

Moffatt HK (1962a) Intensification of the earth's magnetic field by turbulence in the ionosphere. J Geophys Res 67(8):3071–3073

Moffatt HK (1962b) Turbulence in conducting fluids. Mécanique de la Turbulence, Marseille 28 août-2 septembre 1961. Colloq Int Cent Natl Rech Sci 108:395–404

Moffatt HK (2002) G. K. Batchelor and the homogenization of turbulence. Ann Rev Fluid Mech 34:19–35

Moffatt HK (2011) George Batchelor: the post-war renaissance of research in turbulence. In: Davidson PA, Kaneda Y, Moffatt K, Sreenivasan KR (eds) A voyage through turbulence. Cambridge University Press, Cambridge, pp 276–304

Moffatt HK (2012) Homogeneous turbulence: an introductory review. J Turbul 13(1):1–11

Monin AS, Yaglom AM (1971) Statistical fluid mechanics. Mechanics of turbulence, vol 1. MIT Press, Cambridge, MA

Monin AS, Yaglom AM (1975) Statistical fluid mechanics. Mechanics of turbulence, vol 2. MIT Press, Cambridge, MA. Russ. Original 1965

Mouret G (1921) Antoine Chézy. Histoire d'une formule d'hydraulique. Ann ponts chaussées, IIe série LXI:165–269

Munk M (1917) Bericht über Luftwiderstandsmessungen von Streben. Mitteilung 1 der Göttinger Modell-Versuchsanstalt für Aerodynamik. In: Technische Berichte. Herausgegeben von der Flugzeugmeisterei der Inspektion der Fliegertruppen. Heft Nr. 4 (1. Juni 1917), pp 85–96, Tafel XXXX–LXIII

Nebeker F (1995) Calculating the weather: meteorology in the 20th century. Academic Press, San Diego

Newton I (1729) The mathematical principles of natural philosophy. Benjamin Motte, first English translation of the third edition (1726)

Nikuradse J (1926) Untersuchungen über die Geschwindigkeitsverteilung in turbulenten Strömungen. Forsch Geb Ingenieurwesens, 281

Nikuradse J (1933) Experimentelle Untersuchungen zur Turbulenzentstehung. Zeitsch Angew Math Mech (ZAMM) 13(3):174–176

Noether F (1921) Das Turbulenzproblem. Z Angew Math Mech (ZAMM) 1(125–138):218–219

Noether F (1926) Zur asymptotischen Behandlung der stationären Lösungen im Turbulenzproblem. Z Angew Math Mech (ZAMM) 6:232–243

Obukhov AM (1941) On the spectral energy distribution in a turbulent flow. Izv Akad Nauk SSSR Geogr Geofiz 5:453–466

Obukhov AM (1962a) Some specific features of atmospheric turbulence. J Geophys Res 67(8):3011–3014

Obukhov AM (1962b) Some specific features of atmospheric turbulence. J Fluid Mech 13(1):77–81

Oestreich A (2008) Die Deutsche Meteorologische Gesellschaft unter der Leitung von Gustav Hellmann (1907–1923). In: Tetzlaff G, Lüdecke C, Dieter Behr H (eds) 125 Jahre Deutsche Meteorologische Gesellschaft. Ann Meteorol 43:57–65

Onsager L (1945) The distribution of energy in turbulence. Phys Rev 68(11/12):286

Orr WMF (1907) The stability or instability of the steady motions of a perfect liquid and a viscous liquid. Proc R Irish Acad A 27:9–138

Owen PR (1956) Review: the theory of hydrodynamic stability, by C. C. Lin. J Fluid Mech 1(3):349–352

Peppler W (1922) Die Beobachtungen der Marinedrachenstationen Breedene/Meer und St. Michel bei Brügge in den Jahren 1915–1918. Deut Seewarte Aerol Hydrogr Beobachtungen Deut Marine-Stationen während Kriegszeit 1914–1918 4:3–60

Pitot H (1735) Description d'une machine pour mesurer la vitesse des eaux courantes, et le sillage des vaisseaux. In: Histoire de l'Académie royale des sciences, Année MDCCXXXII, avec les

mémoires de mathématique & de physique pour la même Année, tirez des registres de cette Académie, 12 Nov 1732, pp 363–376
Pohlhausen K (1920) Zur näherungsweisen Integration der Differentialgleichung der laminaren Grenzschicht. PhD thesis, Göttingen University, Göttingen
Pohlhausen K (1921) Zur näherungsweisen Integration der Differentialgleichung der laminaren Grenzschicht. ZAMM 1:252–290
Prandtl L (1905) Über Flüssigkeitsbewegung bei sehr kleiner Reibung. In: Verhandlungen des dritten Internationalen Mathematiker-Kongresses in Heidelberg vom 8. bis 13, Aug 1904, pp 484–491
Prandtl L (1910a) Bemerkungen über Dimensionen und Luftwiderstandsformeln. Z Flugtech Motorludtsch 1(13):157–161
Prandtl L (1910b) Eine Beziehung zwischen Wärmeaustausch und Strömungswiderstand der Flüssigkeitsteilchen. Phys Z 11:1072–1078
Prandtl L (1913) Bericht über die Tätigkeit der Göttinger Modellversuchsanstalt. Jahr Motorluftsch-Studiengesellsch 73–81
Prandtl L (1914) Der Luftwiderstand von Kugeln. Nachrichten der Gesellschaft der Wissenschaften zu Göttingen, Mathematisch-physikalische Klasse, pp 177–190
Prandtl L (1921) Bemerkungen über die Entstehung der Turbulenz. Z Angew Math Mech (ZAMM) 1:431–436
Prandtl L (1922) Bemerkungen über die Entstehung der Turbulenz. Phys Z 23:19–25
Prandtl L (1925) Bericht über Untersuchungen zur ausgebildeten Turbulenz. ZAMM 5:136–139
Prandtl L (1926a) Bericht über neuere Turbulenzforschung. In: Hydraulische Probleme. VDI-Verlag, Berlin, pp 1–13. LPGA 2:719–730
Prandtl L (1926b) Zum Turbulenzproblem. Z Angew Math Mech (ZAMM) 6(4):339–340
Prandtl L (1927) Über die ausgebildete Turbulenz. Verhandlungen des II. Internationalen Kongresses für Technische Mechanik. Füssli, Zürich, pp 62–75
Prandtl L (1930) Vortrag in Tokyo. J Aeronaut Res Inst Tokyo Imperial Univ 5(65):12–24
Prandtl L (1932) Zur turbulenten Strömung in Rohren und längs Platten. Ergebnisse Aerodyn Versuchsanstalt Göttingen 4:18–29
Prandtl L (1933) Neuere Ergebnisse der Turbulenzforschung. Z Vereines Deut Ingen 77:105–114
Prandtl L (1948) Mein Weg zu Hydrodynamischen Theorien. Phys Blätter 4:89–92
Prandtl L (1953) Turbulenz. Naturforschung und Medizin in Deutschland 1939–1946. Für Deut bestimmte Ausg FIAT Rev German Sci 11:55–78
Prandtl L, Reichardt H (1934) Einfluss von Wärmeschichtung auf Eigenschaften einer turbulenten Strömung. Deut Forsch 15:110–121
Prandtl L, Tollmien W (1925) Die Windverteilung über dem Erdboden, errechnet aus den Gesetzen der Rohrströmung. Z Geophys 1:47–55
Prandtl L, Wieghardt K (1945) Über ein neues Formelsystem für die ausgebildete Turbulenz. In: Nachrichten der Akademie der Wissenschaften zu Göttingen, Mathematisch-physikalische Klasse, pp 6–19
Rechenberg H (2010) Werner Heisenberg-Die Sprache der Atome. Springer, Berlin, Heidelberg
Redondi P (1985) Along the water: the genius and the theory. Hist Technol 2:77–110
Reichardt H (1933) Die quadratischen Mittelwerte der Längsschwankungen in der turbulenten Kanalströmung. ZAMM 13(3):177–180
Reid EG (1925) Standardization tests of N.A.C.A. No. 1 wind tunnel. NACA Rep 195
Reynolds O (1883) An experimental investigation of the circumstances which determine whether the motion of water in parallel channels shall be direct or sinuous and of the law of resistance in parallel channels. Philos Trans R Soc 174:935–982
Reynolds O (1895) On the dynamical theory of incompressible viscous fluids and the determination of the criterion. Philos Trans R Soc 186A:123–164
Reynolds TS (1983) Stronger than a hundred men: a history of the vertical water wheel. The Johns Hopkins University Press, Baltimore and London
Richardson LF (1920a) Some measurements of atmospheric turbulence. Philos Trans R Soc A221:1–28

Richardson LF (1920b) The supply of energy from and to atmospheric eddies. Proc R Soc A 97:354–373
Richardson LF (1922) Weather prediction by numerical processes. Cambridge University Press, Cambridge
Riegels F (1943) Russische Laminarprofile. Deut Luftfahrtforsch Untersuch Mitteil 3040. ZWB
Robins B (1742) New principles of gunnery. J. Nourse, London
Robins B (1761) An account of the experiments, relating to the resistance of the air, exhibited at different times before the Royal Society, in the year 1746. In: Wilson J (ed) Mathematical tracts of the late Benjamin Robins, vol 1. J. Nourse, London, pp 200–206
Roland A (1985) Model research. The national advisory committee for aeronautics 1915–1958, 2 vols. NASA, Washington, DC. SP-4103
Rollins IM (2017) Sounding the atmosphere during the first world war: American meteorological practice in the signal corps meteorological service, 1917–1937. Master's thesis, Portland State University. University Honors Theses. Paper 405
Rott N (1990) Note on the history of the Reynolds number. Ann Rev Fluid Mech 22:1–11
Rotta JC (1950) Über die Theorie der turbulenten Grenzschichten. Mitteilungen aus dem Max-Planck-Institut für Strömungsforschung. Translated in NACA TM 1344, 1
Rotta JC (1951a) Statistische Theorie nichthomogener Turbulenz. 1. Mitteilung. Z Phys 129:547–572
Rotta JC (1951b) Statistische Theorie nichthomogener Turbulenz. 2. Mitteilung. Z Phys 131:51–77
Rotta JC (1953) On the theory of the turbulent boundary layer. NACA Tech Mem 1344:1–50
Rotta JC (1956) Experimenteller Beitrag zur Entstehung turbulenter Strömung im Rohr. Ingen Arch 24:258–281
Rotta JC (1962) Incompressible turbulent boundary layers. Mécanique de la Turbulence, Marseille 28 août-2 septembre 1961. Colloq Int Cent Natl Rech Sci 108:255–285
Rotta JC (1981) Ein geschichtlicher Rückblick auf die Anfänge der Grenzschichttheorie. DGLR 81–069:1–21
Rouse H, Ince S (1957) History of hydraulics. Iowa Institute of Hydraulic Research, State University of Iowa, Iowa City
Ruckes W (1908) Untersuchungen über den Ausfluss komprimierter Luft aus Kapillaren und die dabei auftretenden Turbulenzerscheinungen. Ann Phys 25:983–1021
Ruden P (1933) Turbulente Ausbreitungsvorgänge im Freistrahl. Naturwissenschaften 23:375–378
Rühlmann M (1885) Vorträge über Geschichte der Technischen Mechanik und Theoretischen Maschinenlehre. Baumgärtner, Leipzig
Saffman PG (1960) On the effect of the molecular diffusivity in turbulent diffusion. J Fluid Mech 8(2):273–283
Saffman PG (1962) Some aspects of the effects of the molecular diffusivity in turbulent diffusion. Mécanique de la Turbulence, Marseille 28 août-2 septembre 1961. Colloq Int Cent Natl Rech Sci 108:53–62
Saffman PG (1992) Vortex dynamics. Cambridge University Press, Cambridge
Saph A, Schoder E (1903) An experimental study of the resistances to the flow of water in pipes. Trans Am Soc Civ Eng 51:253–312
Schaffer S (2004) Fish and ships: models in the age of reason. In: de Chadarevian S, Hopwood N (eds) Models: the third dimension of science. Stanford University Press, Stanford, pp 71–105
Schiehlen W, van Wijngaarden L (eds) (2000) Mechanics at the turn of the century. Shaker Verlag, Aachen
Schiller L (1921) Experimentelle Untersuchungen zum Turbulenzproblem. Z Angew Math Mech (ZAMM) 1:436–444
Schiller L (1934) Neue quantitative Versuche zur Turbulenzentstehung. Z Angew Math Mech (ZAMM) 14(1):36–42
Schlichting H (1930) Über das ebene Windschattenproblem. Ingen-Arch 1:533–571
Schlichting H (1932) Über die Stabilität der Couetteströmung. Ann Phys V 14:505–936

Schlichting H (1933) Zur Entstehung der Turbulenz bei der Plattenströmung. In: Nachrichten von der Gesellschaft der Wissenschaften zu Göttingen, Mathematisch-Physikalische Klasse, pp 181–208
Schlichting H (1941) Der Umschlag laminar/turbulent und das Widerstandsproblem des Tragflügels. Nach einem Vortrag im Seminar der Luftfahrtforschungsanstalt Hermann Göring. In: Deutsche Luftfahrtforschung, Forschungsbericht Nr. 1410. ZWB
Schlichting H (1942) Vortragsreihe 'Grenzschichttheorie' gehalten im Wintersemester 1941/42 in der Luftfahrtforschungsanstalt Hermann Göring. ZWB
Schlichting H (1949a) Lecture series "boundary layer theory", part I: laminar flows. NACA TM No. 1217. Translation of "Vortragsreihe" W.S. 1941/42. Luftfahrtforschungsanstalt Hermann Göring, Braunschweig
Schlichting H (1949b) Lecture series "boundary layer theory", part II: turbulent flows. NACA, TM No. 1218. Translation of "Vortragsreihe" W.S. 1941/42. Luftfahrtforschungsanstalt Hermann Göring, Braunschweig
Schlichting H (1951) Grenzschicht-Theorie. Braun, Karlsruhe. In: völlige Neubearbeitung der 1942 in Typoskriptform erschienenen schriftlichen Ausarbeitung einer 1941/1942 an der Luftfahrtforschungsanstalt in Braunschweig gehalteten Vortragsreihe des Verfassers
Schlichting H, Bußmann K (1942) Messungen an Laminarprofilen. Bericht Lilienthalgesellsch 149:17–20. ZWB
Schlichting H, Gersten K (2017) Boundary-layer theory. Nineth English, Springer, Heidelberg, Berlin
Schmaltz F (2005) Kampfstoff-Forschung im Nationalsozialismus: zur Kooperation von Kaiser-Wilhelm-Instituten. Militär und Industrie, Wallstein, Göttingen
Schmidt W (1917) Der Massenaustausch bei der ungeordneten Strömung in freier Luft und seine Folgen. Kaiserliche Akademie der Wissenschaften in Wien, Mathematisch-naturwissenschaftliche Klasse. Sitzungsberichte Abteilung II a Math Astron Phys Meteorol Mech 126(6):757–804
Schmidt W (1925) Der Massenaustausch in freier Luft und verwandte Erscheinungen. Henri Grand, Hamburg
Schmitt F (2007) About Boussinesq's turbulent viscosity hypothesis: historical remarks and a direct evaluation of its validity. C R Mecanique 335:617–627
Schmitt F (2017) Turbulence from 1870 to 1920: the birth of a noun and of a concept. C R Mecanique 345:620–626
Schubauer GB (1935) A turbulence indicator utilizing the diffusion of heat. NACA Rep 524:337–341
Schubauer GB, Klebanoff PS (1955) Contributions on the mechanism of boundary-layer transition. NACA Rep 1289:853–863
Schubauer GB, Skramstad HK (1943) Laminar-boundary-layer oscillations and transition on a flat plate. NACA advance confidential report. Published in 1948 as NACA report no. 909
Scott JT (2008) Fifty years of physics of fluids. Phys Fluids 20(1):011301-1–011301-4
Sexl T (1927a) Zur Stabilitätsfrage der Poiseuilleschen und Couetteschen Strömung. Ann Phys 83:835–848
Sexl T (1927b) Über dreidimensionale Störungen der Poiseuilleschen Strömung. Ann Phys 84:807–822
Sexl T (1928) "Uber einige Integrale der für die achsensymmetrischen Strömungen in Rohren charakteristischen Differentialgleichung. Ann Phys 87:570–580
Shaw N (1922) Turbulence as a meteorological agency. Nature 109:469–470
Siegmund-Schultze R (2018) Applied mathematics versus fluid dynamics: the catalytic role of Richard von Mises (1883–1953). Hist Stud Nat Sci 48(4):475–525
Siegmund-Schultze R (2020) "The joy that engineers and mathematicians have come together." Richard von Mises' foundation of ZAMM, and its "tasks and goals" (1920/21). ZAMM 100(2):e202002017
Simmons CT (2008) Henry Darcy (1803–1858): immortalised by his scientific legacy. Hydrogeol J 16:1023–1038

Simmons L, Salter C (1934) Experimental investigation and analysis of the velocity variations in turbulent flow. Proc R Soc Lond A 145:212–234

Smeaton J (1759) An experimental enquiry concerning the natural powers of water and wind to turn mills, and other machines, depending on a circular motion. Philos Trans R Soc 51:100–174

Solari G (2019) Wind science and engineering: origins, developments, fundamentals and advancements. Springer

Sommerfeld A (1909) Ein Beitrag zur hydrodynamischen Erklärung der turbulenten Flüssigkeitsbewegung. In: Atti del IV Congresso Internazionale dei Matematici (Roma, 6–11 Aprile 1908), Roma. Tip Della R Accad Lincei 3:116–124

Spalding DB (1991) Kolmogorov's two-equation model of turbulence. Proc Math Phys Sci 434:211–216

Sprat T (1667) The history of the Royal Society of London, for the improving of natural knowledge. Royal Society, London

Sprung A (1885) Lehrbuch der Meteorologie. Hoffmann und Campe, Hamburg

Squire HB (1933) On the stability of three-dimensional disturbances of viscous flow between parallel walls. Proc R Soc A 142:621–628

Sreenivasan KR (2011) G. I. Taylor: the inspiration behind the Cambridge school. In: Davidson PA, Kaneda Y, Moffatt K, Sreenivasan KR (eds) A voyage through turbulence. Cambridge University Press, Cambridge, pp 127–186

Stanton TE, Pannell JR (1914) Similarity of motion in relation to the surface friction of fluids. Philos Trans R Soc A 214:199–224

Steen P, Brutsaert W (2017) Saph and Schoder and the friction law of Blasius. Ann Rev Fluid Mech 49:575–582

Stewart RW, Grant HL (1962) Determination of the rate of dissipation of turbulent energy near the sea surface in the presence of waves. J Geophys Res 67(8):3177–3180

Sutera SP, Skalak R (1993) The history of Poiseuille's law. Ann Rev Fluid Mech 25:1–19

Synge JL (1938) Hydrodynamic stability. Semicent Address Am Math Soc 2:227–269

Synge JL (1939) The stability of plane Poiseuille motion. In: Den Hartog JP, Peters H (eds) Proceedings of the fifth international congress on applied mechanics, Cambridge, MA. Wiley, New York, pp 326–332

Tani I (1977) History of boundary-layer theory. Ann Rev Fluid Mech 9:87–111

Taylor GI (1915) Eddy motion in the atmosphere. Philos Trans R Soc A215:1–26

Taylor GI (1916) Variation of wind velocity close to the ground. Reports and memoranda of the advisory committee for aeronautics, 296(1)

Taylor GI (1917a) Observations and speculations on the nature of turbulent motion. Reports and memoranda of the advisory committee for aeronautics, 345

Taylor GI (1917b) Phenomena connected with turbulence in the lower atmosphere. Proc R Soc A 94:137–155

Taylor GI (1917c) The formation of fog and mist. Q J R Meteorol Soc XLIII:241–268

Taylor GI (1918) On the dissipation of eddies. Reports and memoranda of the advisory committee for aeronautics, 598

Taylor GI (1922) Diffusion by continuous movements. Proc Lond Math Soc 20:196–212

Taylor GI (1923) Stability of a viscous liquid contained between two rotating cylinders. Proc R Soc Lond A 223:289–343

Taylor GI (1932) The transport of vorticity and heat through fluids in turbulent motion. Proc R Soc A 135:685–702

Taylor GI (1935a) Statistical theory of turbulence. Proc R Soc A 151:421–444

Taylor GI (1935b) Statistical theory of turbulence. II. Proc R Soc Lond A 151:444–454

Taylor GI (1935c) Statistical theory of turbulence. III. Distribution of dissipation of energy in a pipe over its cross-section. Proc R Soc Lond A 151:455–464

Taylor GI (1935d) Statistical theory of turbulence. IV. Diffusion in a turbulent air stream. Proc R Soc Lond A 151:465–478

Taylor GI (1936) Statistical theory of turbulence. V. Effect of turbulence on boundary layer. Theoretical discussion of relationship between scale of turbulence and critical resistance of spheres. Proc R Soc Lond A 156:307–317

Taylor GI (1937) The statistical theory of isotropic turbulence. J Aeronaut Sci 4(8):311–315

Taylor GI (1938a) Production and dissipation of vorticity in a turbulent fluid. Proc R Soc 164(916):15–23

Taylor GI (1938b) The spectrum of turbulence. Proc R Soc A 164:476–490

Taylor GI (1939) Some recent developments in the study of turbulence. In: Den Hartog JP, Peters H (eds) Proceedings of the fifth international congress on applied mechanics, Cambridge MA. Wiley, New York, pp 294–310

Taylor GI (1970) Some early ideas about turbulence. J Fluid Mech 41:5–11

Taylor GI, Green AE (1937) Mechanism of the production of small eddies from large ones. Proc R Soc 158(895):499–521

Tennekes H, Lumley JL (1972) A first course in turbulence. MIT Press, Cambridge, MA

Thorpe AJ, Volkert H, Ziemiański MJ (2003) The Bjerknes' circulation theorem: a historical perspective. Bull Am Meteorol Soc 84(4):471–480

Tietjens O (1925) Beiträge zur Entstehung der Turbulenz. Z Angew Math Mech (ZAMM) 5:200–217

Timmermann G (1975) 100 Jahre Experimental-Hydrodynamik im Schiffbau. Deut Schiffahrtsarch 1:135–141

Tollmien W (1926) Berechnung turbulenter Ausbreitungsvorgänge. ZAMM 6:468–478

Tollmien W (1929) Über die Entstehung der Turbulenz. 1. Mitteilung. In: Nachrichten von der Gesellschaft der Wissenschaften zu Göttingen, Mathematisch-Physikalische Klasse, pp 21–44

Tollmien W (1935) Ein allgemeines Kriterium der Instabilität laminarer Geschwindigkeitsverteilungen. In: Nachrichten von der Gesellschaft der Wissenschaften zu Göttingen, Mathematisch-Physikalische Klasse, pp 79–114

Tollmien W (1952) Über Schwingungen in laminaren Strömungen und die Theorie der Turbulenz. In: Proceedings of the 8th international congress for theoretical and applied mechanics, Istanbul, 20–28 Aug 1952, pp 23–47

Tollmien W (1953) Zusammenfassender Bericht. Fortschritte der Turbulenzforschung. ZAMM 33(5/6):200–211

Townsend AA (1956) The structure of turbulent shear flow. Cambridge University Press, Cambridge

Truesdell C (1954) Rational fluid mechanics, 1687–1765. In: Editor's introduction to Euleri Opera Omnia, series II, vol 12. Zurich, Fuessli, pp IX–CXXV

Tulapurkara EG (2005) Hundred years of the boundary layer-some aspects. Sadhana 30(4):499–512

Turnbull D (1995) Rendering turbulence orderly. Soc Stud Sci 25:9–33

Uberoi MS, Corrsin S (1953) Diffusion of heat from a line source in isotropic turbulence. NACA Rep 1142:1–29

Volkert H (1999) Components of the Norwegian cyclone model: observations and theoretical ideas in Europe prior to 1920. In: Shapiro MA, Gronas S (eds) The life cycles of extratropical cyclones. American Meteorological Society, Boston, pp 15–28

von Helmholtz H (1865) Populäre wissenschaftliche Vorträge. Vieweg, Braunschweig

von Helmholtz H (1888) Über atmosphärische Bewegungen. In: Hermann von Helmholtz: Wissenschaftliche Abhandlungen. Barth, 1882–1895, Leipzig, pp 289–308

von Kármán T (1911) Über den Mechanismus des Widerstandes, den ein bewegter Körper in einer Flüssigkeit erfährt. In: Nachrichten der K. Gesellschaft der Wissenschaften zu Göttingen, mathematisch-physikalische Klasse, pp 509–517

von Kármán T (1912) Über den Mechanismus des Widerstandes, den ein bewegter Körper in einer Flüssigkeit erfährt. In: Nachrichten der K. Gesellschaft der Wissenschaften, mathematisch-physikalische Klasse, pp 547–556

von Kármán T (1921) Über laminare und turbulente Reibung. ZAMM 1:233–252

von Kármán T (1922) Über die Oberflächenreibung von Flüssigkeiten. In: von Kármán T, Leci-Civita T (Hrsg) Vorträge aus dem Gebiete der Hydro- und Aerodynamik (Innsbruck 1922). Springer, Berlin, pp 146–167. CWTK 2:133–152

von Kármán T (1930) Mechanische Ähnlichkeit und Turbulenz. In: Nachrichten von der Gesellschaft der Wissenschaften zu Göttingen, Mathematisch-Physikalische Klasse, pp 58–76
von Kármán T (1931) Mechanische Ähnlichkeit und Turbulenz. In: Oseen ACW, Weibull W (eds) Proceedings of the third international congress of applied mechanics, 24–29 Aug 1930, 3 vols. AB. Sveriges Litografiska Tryckerier, Stockholm, 1, pp 85–93
von Kármán T (1934) Turbulence and skin friction. J Aeronaut Sci 1:1–20
von Kármán T (1937a) The fundamentals of the statistical theory of turbulence. J Aeronaut Sci 4(4):131–138
von Kármán T (1937b) Turbulence. J R Aeronaut Soc 41:1108–1141. In: Twenty-fifth Wilbur Wright memorial lecture delivered to the Royal Aeronautical Society on May 27th, 1937
von Kármán T (1967) The wind and beyond. Little, Brown and Company, Boston/Toronto (with Lee Edson)
von Kármán T, Howarth L (1938) On the statistical theory of isotropic turbulence. Proc R Soc 164:192–215
von Kármán T, Levi-Civita T (1924) Vorträge aus dem Gebiete der Hydro- und Aerodynamik (Innsbruck 1922). Springer, Berlin
von Mises R (1914) Elemente der technischen Hydromechanik. Teubner, Leipzig
von Mises R (1921a) Zur Einführung. Über die Aufgaben und Ziele der angewandten Mathematik. Z Angew Math Mech (ZAMM) 1:1–15
von Mises R (1921b) Zur Versammlung der Mathematiker und Physiker in Jena. Z Angew Math Mech (ZAMM) 1:341–342
von Weizsäcker CF (1943) Über die Entstehung des Planetensystems. Z Astrophys 22:319–355
von Weizsäcker CF (1948) Das Spektrum der Turbulenz bei großen Reynoldsschen Zahlen. Z Phys 124:614–627
Wali KC (1991) Chandra: a biography of S. Chandrasekhar. University of Chicago Press, Chicago, IL
Wallace J (2013) Highlights from 50 years of turbulent boundary layer research. J Turbul 13(53):1–70
Walz A (1962) Compressible turbulent boundary layers. Mécanique de la Turbulence, Marseille 28 août-2 septembre 1961. Colloq Int Cent Natl Rech Sci 108:299–350
Ward RD (1915) The weather factor in the great war: late winter and spring. J Geogr 13(10):315–317
Ward RD (1916) The weather factor in the great war: IV: October, 1915–February, 1916. J Geogr 14(10):373–384
Wege K (1996) Zur Historie des Meteorologischen Observatoriums Hohenpeißenberg. Promet 25(4):90–98
Weisbach J (1845) Lehrbuch der Ingenieur- und Maschinen-Mechanik, vol 1. Vieweg, Braunschweig
Weisskopf VF (1954) Visas for foreign scientists. Bull Atom Sci 10:68–80
Wesfreid JE (2017) Henri Bénard: thermal convection and vortex shedding. C R Mecanique 345:446–466
Wieghardt K (1944) Zum Reibungswiderstand rauher Platten. Deut Luftfahrtforsch Untersuch Mitteil 6612. ZWB
Wien W (1900) Lehrbuch der Hydrodynamik. Hirzel, Leipzig
Wieselsberger C (1914) Der Luftwiderstand von Kugeln. Z Flugtech Motorluftsch 5:140–145
Woltmann R (1791) Beiträge zur hydraulischen Architektur, vol 1. Dieterich, Göttingen
Yaglom AM (1962) Some mathematical models generalizing the model of homogeneous and isotropic turbulence. J Geophys Res 67(8):3081–3087
Yaglom AM (1990) Obituary: Alexander Mikhailovich Obukhov, 1918–1989. Boundary-Layer Meteorology 53:v–xi
Yaglom AM (1994) A. N. Kolmogorov as a fluid mechanician and founder of a school in turbulence research. Ann Rev Fluid Mech 26:1–22

Index

M. Eckert, *Turbulence—an Odyssey*, History of Physics,
https://doi.org/10.1007/978-3-030-91459-2